Pitman Research Notes in Mathematics Series

Submission of proposals for consideration
Suggestions for publication, in the form of outlines and representative samples, are invited by the Editorial Board for assessment. Intending authors should approach one of the main editors or another member of the Editorial Board, citing the relevant AMS subject classifications. Alternatively, outlines may be sent directly to the publisher's offices. Refereeing is by members of the board and other mathematical authorities in the topic concerned, throughout the world.

Preparation of accepted manuscripts
On acceptance of a proposal, the publisher will supply full instructions for the preparation of manuscripts in a form suitable for direct photo-lithographic reproduction. Specially printed grid sheets can be provided and a contribution is offered by the publisher towards the cost of typing. Word processor output, subject to the publisher's approval, is also acceptable.

Illustrations should be prepared by the authors, ready for direct reproduction without further improvement. The use of hand-drawn symbols should be avoided wherever possible, in order to maintain maximum clarity of the text.

The publisher will be pleased to give any guidance necessary during the preparation of a typescript, and will be happy to answer any queries.

Important note
In order to avoid later retyping, intending authors are strongly urged not to begin final preparation of a typescript before receiving the publisher's guidelines. In this way it is hoped to preserve the uniform appearance of the series.

Addison Wesley Longman Ltd
Edinburgh Gate
Harlow, Essex, CM20 2JE
UK
(Telephone (0) 1279 623623)

Titles in this series. A full list is available from the publisher on request.

B N Mandal and Nanigopal Mandal

Indian Statistical Institute, Calcutta

Integral expansions related to Mehler–Fock type transforms

Some new types of integral transforms involving spherical harmonics

CRC Press

Taylor & Francis Group

Boca Raton London New York

CRC Press is an imprint of the
Taylor & Francis Group, an **informa** business

A CHAPMAN & HALL BOOK

CRC Press
Taylor & Francis Group
6000 Broken Sound Parkway NW, Suite 300
Boca Raton, FL 33487-2742

© 1997 by Taylor & Francis Group, LLC
CRC Press is an imprint of Taylor & Francis Group, an Informa business

First issued in paperback 2019

No claim to original U.S. Government works

ISBN 13: 978-0-367-44830-1 (pbk)
ISBN 13: 978-0-582-30816-9 (hbk)

Visit the Taylor & Francis Web site at
http://www.taylorandfrancis.com

and the CRC Press Web site at
http://www.crcpress.com

Contents

Contents

Preface

The aim of this book is to present in a systematic manner various integral expansions related to the *Mehler–Fock* and *Mehler–Fock type transforms* involving spherical harmonics, namely, Legendre, associated Legendre, generalized associated Legendre, non-periodic Legendre and associated Legendre functions. In the classical Mehler–Fock transform formulae, the kernel is the Legendre function $P_\nu(z)$ or associated Legendre function $P_\nu^\mu(z)$ where μ is zero or integer, $\nu = -\frac{1}{2} + i\tau$, τ is a real parameter with $z > 1$ and in the inverse transform formulae, the subscript ν appears as an integration variable. There is another class of integral transform formulae involving associated Legendre functions somewhat related to the Mehler–Fock transforms in which the superscript of the associated Legendre functions appears as the integration variable in the inverse transform formula while the subscript remains fixed. This class of integral transforms may be termed *Mehler–Fock type transforms*. This nomenclature is used here to distinguish these from the classical *Mehler–Fock transforms*. These are not widely known in the literature although the first paper in this area was published in 1958 by L. B. Felsen. This is perhaps due to the non-availability of any book or monograph dealing with these integral expansions. This book will perhaps remove this lacuna and attract mathematicians to this area so as to enrich it further.

It is felt that most of this book will be accessible to a reader well grounded in the methods of advanced calculus and in the theory of complex analysis.

The book contains five chapters. Chapter 1 is confined to the properties of the spherical harmonics which are not easily available in any book or monograph. Chapter 2 is devoted to the establishment, together with their proof, of some integral expansions related to the Mehler–Fock transforms. In Chapter 3, some integral expansions related to Mehler–Fock type transforms involving associated Legendre functions are established by presenting rigorous proofs in the classical manner. Chapter 4 is concerned with integral expansion formulae related to the Mehler—Fock type transforms involving generalized associated Legendre functions. Lastly, some properties of the non-periodic Legendre and associated Legendre functions together with the integral expansions involving these functions are discussed in Chapter 5. Illustrative examples for these integral expansions are also given in the form of integral expansions of some simple functions. It is hoped that this book will stimulate some mathematicians to search for applications of these integral expansions to appropriate problems of mathematical physics.

We appreciate the wisdom of Professor S. C. Dasgupta, Retired Professor of Mathematics, Bengal Engineering College, who introduced one of us to this area almost three decades back. The book was completed during the tenure of a Research Associateship offered by the Council of Scientific and Industrial Research, New Delhi, to NM. We would like to thank Dr. D. C. Dalal, Dr. D. P. Dolai and Dr. M. Kanoria for their suggestions and help in processing the manuscript. Finally, we thank Professor A. Jeffrey for his encouragement and the staff of Addison Wesley Longman for their cooperation in bringing out this publication.

December, 1996   B. N. Mandal
Indian Statistical Institute   Nanigopal Mandal
Calcutta, India

Chapter 1

Introduction

Any natural phenomenon has to be modelled mathematically if one wants to study it on a sound scientific footing. In many situations mathematical modelling of physical phenomena gives rise to boundary value problems or initial value problems wherein one has to solve a partial differential equation with prescribed values at boundaries. There has been a rich development of various mathematical techniques for the solution of boundary value problems or initial value problems in the literature. The various techniques of applied mathematics play very significant roles in the study of these problems. The technique of integral transforms is a powerful and indispensible tool for the modern applied mathematician or theoretical physicist for successful handling of boundary value problems arising in mathematical physics.

An important class of integral expansions generated by Sturm–Liouville theory involving spherical harmonics is commonly known as *Mehler–Fock transforms* (cf. Lebedev, 1965, Ch8 and Sneddon, 1972, Ch7). The classical Mehler-Fock transform has been successfully applied to deal with problems occurring in the mathematical theory of elasticity, particularly those related to the use of toroidal and ellipsoidal coordinates. In these transform formulae, the kernel is the Legendre function $P_\nu(z)$ or associated Legendre function $P_\nu^\mu(z)$ or generalized associated Legendre function $P_\nu^{m,n}(z)$ (where $\nu = -\frac{1}{2}+i\tau, \tau$ is a real parameter) with $z > 1$ and in the inverse transform formula, the subscript ν of the spherical harmonics appears as the integration variable.

There is another class of integral expansions involving associated Legendre and generalized associated Legendre functions somewhat related to Mehler–Fock transforms, in which the superscript of the associated spherical harmonics appears as the integration variable in the inverse transform formula while the subscript remains fixed. This class of integral expansions may be termed *Mehler–Fock type transforms*.

Formal construction of an integral transform together with its inverse formula is somewhat straightforward because of the known integral representation of the Dirac δ-function in terms of the Green's function of a second order Sturm–Liouville type linear differential equation (cf. Marcuvitz, 1951, Friedman, 1956, p 218, Clemmow, 1961). However, rigorous proof for the establishment of an integral transform and its inverse formula is not at all easy. The formal method of construction is based on the following somewhat intuitive argument.

Let a second order Sturm–Liouville type homogeneous differential equation be

$$-\frac{d}{dx}\left\{p(x)\frac{du}{dx}\right\} + \{q(x) - \lambda\, r(x)\}\, u = 0, \quad a < x < b, \tag{1.1}$$

where $p(x), q(x), r(x)$ are piecewise continuous in (a, b) and u satisfies specified boundary conditions at the end points. The Green's function $G(x, \xi, \lambda)$ satisfies the inhomogeneous differential equation

$$\left\{-\frac{d}{dx}\left(p(x)\frac{d}{dx}\right) + q(x) - \lambda\, r(x)\right\} G(x, \xi, \lambda) = \delta(x - \xi), \tag{1.2}$$

and is given by

$$G(x, \xi, \lambda) = \begin{cases} \dfrac{u_1(\xi)\, u_2(x)}{p\,W}, & x < \xi, \\[3mm] \dfrac{u_1(x)\, u_2(\xi)}{p\,W}, & x > \xi, \end{cases} \tag{1.3}$$

where $u_1(x), u_2(x)$ are two independent solutions of the differential equation (1.1), and

$$W(x) = u_1(x)\, u_2{}'(x) - u_1{}'(x)\, u_2(x)$$

is their Wronskian, so that $p(x)\,W(x)$ is a constant.

A unique δ-function representation of the Green's function $G(x, \xi, \lambda)$ is given by

$$\frac{\delta(x - \xi)}{r(\xi)} = -\frac{1}{2\pi i}\oint G(x, \xi, \lambda)\, d\lambda, \tag{1.4}$$

2

where the path of integration is a contour enclosing anticlockwise all the singularities of $G(x, \xi, \lambda)$ in the complex λ-plane. The construction of the Green's function depends critically on the given boundary conditions. Now any function $f(x)$ defined in $a < x < b$ can formally be represented by

$$f(x) = \int_a^b \delta(x - \xi) \, f(\xi) \, d\xi = \frac{1}{2\pi i} \oint \frac{u_1(x)}{p \, W} \left[\int_a^b r(\xi) \, f(\xi) \, u_2(\xi) \, d\xi \right] d\lambda. \qquad (1.5)$$

The integral inside the square bracket in equation (1.5) gives the integral transform while the whole of the right hand side gives the integral expansion formula for the function $f(x)$. It is noted that, although $G(x, \xi, \lambda)$ is different for $x < \xi$ and $x > \xi$, the representation of $\delta(x - \xi)$ is the same for both $x < \xi$ and $x > \xi$. The key to obtaining formally the integral transform and the corresponding inversion formula lies in this fact. This idea will be utilized in Sections 3.3. and 5.1 to obtain formal integral expansion of functions in terms of Legendre functions.

The aims of this book are to present in a systematic manner various integral expansions related to the aforesaid Mehler–Fock and Mehler–Fock type transforms involving spherical harmonics. It may be noted that these integral expansions can be viewed as expansions in eigenfunctions of the Sturm–Liouville type differential equation

$$\left((1 - x^2) \, u' \right)' - \left(\frac{m^2}{2(1 - x)} + \frac{n^2}{2(1 + x)} + \frac{1}{4} + \tau^2 \right) u = 0, \qquad (1.6)$$

$x \in (-1, 1)$ or $(1, \infty)$ with first and second kind boundary conditions at the ends (here m, n are zero or integers or complex numbers, and τ is a real parameter). These integral expansions are justified for sufficiently smooth functions and the subsequent analysis presented here is classical in nature.

In this chapter, we describe some properties of spherical harmonics which will be useful in the presentation of the mathematical analysis given in the subsequent chapters. Detailed properties are available in the literature (cf. Erdélyi et al. 1953, Hobson, 1931). We confine ourselves to some results mostly needed in the book.

1.1 Some properties of Legendre functions

The functions $P_\nu(z)$ and $Q_\nu(z)$, two specific linearly independent solutions of the differential equation

$$(1 - z^2)\frac{d^2u}{dz^2} - 2z\frac{du}{dz} + \nu\,(\nu + 1)\,u = 0, \tag{1.1.1}$$

ν, z being unrestricted, are known as Legendre functions of the first and second kind respectively.

The functions $P_\nu(z)$ and $Q_\nu(z)$ can be expressed, in terms of the hypergeometric functions, in the following forms (cf. Erdélyi et al. 1953, p 142(21) and p 137(44)):

$$P_\nu(z) = (2\pi)^{-\frac{1}{2}}\,(z^2 - 1)^{-\frac{1}{4}}\frac{\Gamma(1+\nu)}{\Gamma(\nu+\frac{3}{2})}\times$$

$$\times\left\{(z + \sqrt{z^2 - 1})^{\nu+\frac{1}{2}}\;F\left(\frac{1}{2},\frac{1}{2};\nu + \frac{3}{2};\frac{z + \sqrt{z^2 - 1}}{2\sqrt{z^2 - 1}}\right)\right.$$

$$\left.+i(z - \sqrt{z^2 - 1})^{\nu+\frac{1}{2}}\;F\left(\frac{1}{2},\frac{1}{2};\nu + \frac{3}{2};\frac{-z + \sqrt{z^2 - 1}}{2\sqrt{z^2 - 1}}\right)\right\}, \tag{1.1.2}$$

and

$$Q_\nu(z) = \sqrt{\frac{\pi}{2}}\,\frac{\Gamma(1+\nu)}{\Gamma(\nu+\frac{3}{2})}\,\frac{(z - \sqrt{z^2 - 1})^{\nu+\frac{1}{2}}}{(z^2 - 1)^{\frac{1}{4}}}\;F\left(\frac{1}{2},\frac{1}{2};\nu + \frac{3}{2};\frac{-z + \sqrt{z^2 - 1}}{2\sqrt{z^2 - 1}}\right), \tag{1.1.3}$$

where the hypergeometric function $F\,(a, b; c; z)$ represents the series

$$F(a, b; c; z) = \sum_{k=0}^{\infty}\frac{(a)_k\,(b)_k}{k!\,(c)_k}\,z^k, \quad (|z| < 1), \tag{1.1.4}$$

with

$$(\lambda)_k = \lambda(\lambda + 1)\ldots(\lambda + k - 1), \quad (\lambda)_0 = 1.$$

Using the above two representations, for any complex number ν, it follows that (cf. Lebedev and Skal'skaya, 1966a)

$$\frac{P_{\nu-\frac{1}{2}}(i\sinh\alpha) + P_{\nu-\frac{1}{2}}(-i\sinh\alpha)}{2} = \frac{\Gamma(\frac{1}{2}+\nu)\cos(\frac{\pi\nu}{2} - \frac{1}{4}\pi)}{(2\pi)^{\frac{1}{2}}\Gamma(1+\nu)\sqrt{\cosh\alpha}}\times$$

$$\times\left\{\frac{F\left(\frac{1}{2},\frac{1}{2};1 + \nu;\frac{e^\alpha}{2\cosh\alpha}\right)}{e^{-\alpha\nu}} + \frac{F\left(\frac{1}{2},\frac{1}{2};1 + \nu;\frac{e^{-\alpha}}{2\cosh\alpha}\right)}{e^{\alpha\nu}}\right\}, \quad -\infty < \alpha < \infty, \tag{1.1.5}$$

4

$$\frac{P_{\nu-\frac{1}{2}}(i\sinh\alpha) - P_{\nu-\frac{1}{2}}(-i\sinh\alpha)}{2i} = \frac{\Gamma(\frac{1}{2}+\nu)\sin(\frac{\pi\nu}{2} - \frac{1}{4}\pi)}{(2\pi)^{\frac{1}{2}}\Gamma(1+\nu)\sqrt{\cosh\alpha}} \times$$

$$\times \left\{ \frac{F\left(\frac{1}{2},\frac{1}{2};1+\nu;\frac{e^{\alpha}}{2\cosh\alpha}\right)}{e^{-\alpha\nu}} - \frac{F\left(\frac{1}{2},\frac{1}{2};1+\nu;\frac{e^{-\alpha}}{2\cosh\alpha}\right)}{e^{\alpha\nu}} \right\}, \quad -\infty < \alpha < \infty, \qquad (1.1.6)$$

$$\frac{Q_{\nu-\frac{1}{2}}(i\sinh\alpha) + Q_{\nu-\frac{1}{2}}(-i\sinh\alpha)}{2} = -\left(\frac{\pi}{2}\right)^{\frac{1}{2}}\frac{\Gamma(\frac{1}{2}+\nu)\sin(\frac{\pi\nu}{2} - \frac{1}{4}\pi)}{\Gamma(1+\nu)\sqrt{\cosh\alpha}}e^{-\alpha\nu} \times$$

$$\times F\left(\frac{1}{2},\frac{1}{2};1+\nu;\frac{e^{-\alpha}}{2\cosh\alpha}\right), \quad \alpha \geq 0, \nu \neq -2n - \frac{1}{2}, n = 0,1,2,\ldots, \qquad (1.1.7)$$

$$\frac{Q_{\nu-\frac{1}{2}}(i\sinh\alpha) - Q_{\nu-\frac{1}{2}}(-i\sinh\alpha)}{2i} = -\left(\frac{\pi}{2}\right)^{\frac{1}{2}}\frac{\Gamma(\frac{1}{2}+\nu)\cos(\frac{\pi\nu}{2} - \frac{1}{4}\pi)}{\Gamma(1+\nu)\sqrt{\cosh\alpha}}e^{-\alpha\nu} \times$$

$$\times F\left(\frac{1}{2},\frac{1}{2};1+\nu;\frac{e^{-\alpha}}{2\cosh\alpha}\right), \quad \alpha \geq 0, \nu \neq -2n - \frac{3}{2}, n = 0,1,2,\ldots. \qquad (1.1.8)$$

Estimates and asymptotic expansions of Legendre functions

From the representations (1.1.5) and (1.1.6), together with the inequality

$$\left| F\left(\frac{1}{2},\frac{1}{2};1+i\tau;x\right) \right| \leq F\left(\frac{1}{2},\frac{1}{2};1;x\right), \quad 0 \leq \tau < \infty, 0 \leq x < 1, \qquad (1.1.9)$$

the following estimates can be obtained (cf. Lebedev and Skal'skaya, 1966a):

$$\left| \frac{P_{-\frac{1}{2}+i\tau}(ix) + P_{-\frac{1}{2}+i\tau}(-ix)}{2} \right| \leq \left(\frac{\sinh\pi\tau}{\pi\tau}\right)^{\frac{1}{2}} \frac{P_{-\frac{1}{2}}(ix) + P_{-\frac{1}{2}}(-ix)}{2}, \qquad (1.1.10)$$

$$\left| \frac{P_{-\frac{1}{2}+i\tau}(ix) - P_{-\frac{1}{2}+i\tau}(-ix)}{2i} \right| < \left(\frac{\sinh\pi\tau}{\pi\tau}\right)^{\frac{1}{2}} \frac{P_{-\frac{1}{2}}(ix) + P_{-\frac{1}{2}}(-ix)}{2}, \qquad (1.1.11)$$

where $0 \leq \tau < \infty$ and $-\infty < x < \infty$.

To obtain the asymptotic expansions of Legendre functions as $|\nu| \to \infty$ with $|\arg\nu| < \frac{\pi}{2}$, we shall make use of the fact that every hypergeometric function in the relations (1.1.5)–(1.1.8) can be expressed in the form (cf. Lebedev and Skal'skaya, 1966a)

$$F\left(\frac{1}{2},\frac{1}{2};1+\nu;x\right) = 1 + \sum_{k=1}^{\infty}\frac{(\frac{1}{2})_k(\frac{1}{2})_k}{(1+\nu)_k k!}x^k$$

$$= 1 + \frac{x}{4(1+\nu)}\sum_{k=0}^{\infty}\frac{(\frac{3}{2})_k(\frac{3}{2})_k}{(2+\nu)_k(k+1)!}x^k = 1 + r(\nu,x), \quad \text{say}, \quad (|x| < 1).$$

Using the inequality

$$\left|\frac{(\frac{3}{2})_k}{(2+\nu)_k}\right|_{k=0,1,2,\ldots} \le 1, \quad |\arg \nu| \le \frac{\pi}{2},$$

the estimate of $r(\nu,x)$ can be obtained as

$$|r(\nu,x)| \le \frac{x}{4|(1+\nu)|}\sum_{k=0}^{\infty}\frac{(\frac{3}{2})_k}{k!}x^k = \frac{x}{4|(1+\nu)|}(1-x)^{-\frac{3}{2}}.$$

Therefore, we can write

$$F\left(\frac{1}{2},\frac{1}{2};1+\nu;\frac{e^{\pm\alpha}}{2\cosh\alpha}\right) = 1 + e^{\pm 2\alpha}\left(e^{\pm 2\alpha}+1\right)^{\frac{1}{2}}O(|\nu|^{-1}), \tag{1.1.12}$$

as $|\nu| \to \infty$, $|\arg \nu| \le \frac{\pi}{2}$, where O is independent of α ($\alpha \ge 0$).

Using the estimate (1.1.12), together with known asymptotic expansions for the gamma functions, it can be obtained that as $|\nu| \to \infty$ with $|\arg \nu| \le \frac{\pi}{2}$ (cf. Belova and Ufliand, 1970, Lebedev and Skal'skaya, 1966a)

$$P_{\nu-\frac{1}{2}}(\cosh\alpha) = \left(\frac{1}{2\pi\nu\sinh\alpha}\right)^{\frac{1}{2}}\times$$

$$\times\left\{e^{\nu\alpha}[1+\sqrt{\cosh\alpha}\,O(|\nu|^{-1})] \pm ie^{-\nu\alpha}[1+O(|\nu|^{-1})]\right\}, \quad \alpha > 0, \tag{1.1.13}$$

$$Q_{\nu-\frac{1}{2}}(\cosh\alpha) = \left(\frac{\pi}{2\nu\sinh\alpha}\right)^{\frac{1}{2}}e^{-\nu\alpha}[1+O(|\nu|^{-1})], \quad \alpha > 0, \tag{1.1.14}$$

$$\frac{P_{\nu-\frac{1}{2}}(i\sinh\alpha)+P_{\nu-\frac{1}{2}}(-i\sinh\alpha)}{2P_{\nu-\frac{1}{2}}(0)} = \frac{1}{2\sqrt{\cosh\alpha}}\times$$

$$\times\left\{\frac{[1+(e^{2\alpha}+1)^{\frac{3}{2}}O(|\nu|^{-1})]}{e^{-\alpha\nu}}+\frac{[1+(e^{-2\alpha}+1)^{\frac{3}{2}}O(|\nu|^{-1})]}{e^{\alpha\nu}}\right\}, \quad -\infty < \alpha < \infty, \tag{1.1.15}$$

$$\frac{\pi\nu P_{\nu-\frac{1}{2}}(0)}{\cos\pi\nu}\frac{P_{\nu-\frac{1}{2}}(i\sinh\alpha)-P_{\nu-\frac{1}{2}}(-i\sinh\alpha)}{2i} = \frac{1}{2\sqrt{\cosh\alpha}}\times$$

$$\times\left\{\frac{[1+(e^{-2\alpha}+1)^{\frac{3}{2}}O(|\nu|^{-1})]}{e^{\alpha\nu}}-\frac{[1+(e^{2\alpha}+1)^{\frac{3}{2}}O(|\nu|^{-1})]}{e^{-\alpha\nu}}\right\}, \quad -\infty < \alpha < \infty, \tag{1.1.16}$$

$$\frac{\nu P_{\nu-\frac{1}{2}}(0)}{\cos\pi\nu}\frac{Q_{\nu-\frac{1}{2}}(i\sinh\alpha)+Q_{\nu-\frac{1}{2}}(-i\sinh\alpha)}{2}$$

$$= \frac{e^{-\alpha\nu}}{2\sqrt{\cosh\alpha}}\left[1+(e^{-2\alpha}+1)^{\frac{3}{2}}O(|\nu|^{-1})\right], \quad \alpha \ge 0, \tag{1.1.17}$$

$$\frac{Q_{\nu-\frac{1}{2}}(i\sinh\alpha) - Q_{\nu-\frac{1}{2}}(-i\sinh\alpha)}{2\pi\,i\,P_{\nu-\frac{1}{2}}(0)} = \frac{-e^{-\alpha\nu}}{2\sqrt{\cosh\alpha}}\left[1 + (e^{-2\alpha}+1)^{\frac{3}{2}}O(|\nu|^{-1})\right],\quad \alpha \geq 0.$$

$$(1.1.18)$$

In the above estimates $(1.1.15)-(1.1.18)$,

$$P_{\nu-\frac{1}{2}}(0) = \frac{\cos\pi\nu}{2\pi\sqrt{\pi}}\,\Gamma\left(\frac{1}{4}+\frac{\nu}{2}\right)\,\Gamma\left(\frac{1}{4}-\frac{\nu}{2}\right)$$

is introduced in order to present the formulae in a more symmetrical form.

The following estimate and identity (cf. Lebedev and Skal'skaya, 1966a) are also useful in our discussion:

$$\frac{P_{-\frac{1}{2}}(ix) + P_{-\frac{1}{2}}(-ix)}{2} = \begin{cases} O(1), & x \in (0,a),\ a > 0, \\ O(1)\,x^{-\frac{1}{2}}\,\ln(1+x), & x \in (a,\infty), \end{cases}$$

$$(1.1.19)$$

and

$$\frac{P_{-\frac{1}{2}+i\tau}(ix) + P_{-\frac{1}{2}+i\tau}(-ix)}{2} = \frac{-\dfrac{d}{dx}\left\{\dfrac{P^1_{-\frac{1}{2}+i\tau}(ix) + P^1_{-\frac{1}{2}+i\tau}(-ix)}{2(x^2+1)^{-\frac{1}{2}}}\right\}}{\frac{1}{4}+\tau^2},\quad -\infty < x < \infty.$$

$$(1.1.20)$$

Introducing the following temporary notation

$$G_\nu(x,y) \equiv P_{\nu-\frac{1}{2}}(ix)P_{\nu-\frac{1}{2}}(-iy) - P_{\nu-\frac{1}{2}}(iy)P_{\nu-\frac{1}{2}}(-ix)$$

$$= \frac{2\cos\pi\nu}{\pi}\left[P_{\nu-\frac{1}{2}}(ix)Q_{\nu-\frac{1}{2}}(iy) - P_{\nu-\frac{1}{2}}(iy)Q_{\nu-\frac{1}{2}}(ix)\right],\quad (1.1.21)$$

and

$$E(x) \equiv e^{\nu x}\,F\left(\frac{1}{2},\frac{1}{2};1+\nu;\frac{e^z}{2\cosh x}\right),\quad (1.1.22)$$

where $-\infty < x < \infty, -\infty < y < \infty$, ν is any complex number, it is found that (cf. Moshinskii, 1989)

$$G_\nu(\sinh\alpha,\sinh\beta) = \frac{i\,\Gamma^2(\nu+\frac{1}{2})\cos\pi\nu\,[E(-\alpha)\,E(\beta) - E(\alpha)\,E(-\beta)]}{\pi\,\Gamma^2(1+\nu)(\cosh\alpha\cosh\beta)^{\frac{1}{2}}},\quad (1.1.23)$$

and

$$\frac{Q_{\nu-\frac{1}{2}}(i\sinh\alpha)}{Q_{\nu-\frac{1}{2}}(i\sinh\beta)} = \left(\frac{\cosh\beta}{\cosh\alpha}\right)^{\frac{1}{2}}\frac{E(-\alpha)}{E(-\beta)},\quad -\infty < \alpha < \infty, -\infty < \beta < \infty.\quad (1.1.24)$$

Using the inequality (1.1.9), we obtain (cf. Moshinskii, 1989) that

$$|G_{i\tau}(x,y)| \leq \frac{\sinh \pi \tau}{\pi \tau} \left[P_{-\frac{1}{2}}(ix) \, P_{-\frac{1}{2}}(-iy) + P_{-\frac{1}{2}}(-ix) \, P_{-\frac{1}{2}}(iy) \right]$$

$$= \frac{\sinh \pi \tau}{\pi \tau} \begin{cases} O(1), & x \in (y,a), \quad a > 0, \\ O(1) \; x^{-\frac{1}{2}} \ln x, & x \in (a,\infty). \end{cases} \tag{1.1.25}$$

The asymptotic expansions of the expressions (1.1.23) and (1.1.24), as $|\nu| \to \infty$ with $|\arg \nu| \leq \frac{\pi}{2}$, can be expressed as (cf. Moshinskii, 1989)

$$G_\nu(\sinh \alpha, \sinh \beta) = \frac{i \cos \pi \nu \, [1 + O(|\nu|^{-1})]}{\sqrt{\cosh \alpha \cosh \beta}} \{\exp[\nu(\beta - \alpha)] \times$$

$$\times [1 + \exp(2\beta)\sqrt{1 + \exp(2\beta)} \, O(|\nu|^{-1}) + \exp(-2\alpha)\sqrt{1 + \exp(-2\alpha)} \, O(|\nu|^{-1})]$$

$$- \exp[\nu(\alpha - \beta)][1 + \exp(2\alpha)\sqrt{1 + \exp(2\alpha)} \, O(|\nu|^{-1})$$

$$+ \exp(-2\beta)\sqrt{1 + \exp(-2\beta)} \, O(|\nu|^{-1})]\} \tag{1.1.26}$$

and

$$\frac{Q_{\nu-\frac{1}{2}}(i \sinh \alpha)}{Q_{\nu-\frac{1}{2}}(i \sinh \beta)} = \sqrt{\frac{\cosh \beta}{\cosh \alpha}} \exp[\nu(\alpha - \beta)] \left\{ 1 + \exp(-2\alpha)\sqrt{1 + \exp(-2\alpha)} O(|\nu|^{-1}) \right.$$

$$\left. + \exp(-2\beta)\sqrt{1 + \exp(-2\beta)} O(|\nu|^{-1}) \right\}, \tag{1.1.27}$$

where $\alpha, \beta \in (-\infty, \infty)$.

1.2 Some properties of associated Legendre functions

The solutions of the differential equation

$$(1 - z^2)\frac{d^2 u}{dz^2} - 2z\frac{du}{dz} + \left[\nu(\nu + 1) - \frac{\mu^2}{1 - z^2} \right] u = 0, \tag{1.2.1}$$

are $P_\nu^\mu(z)$ and $Q_\nu^\mu(z)$ and are known as associated Legendre functions of the first and second kind respectively. For $\mu = 0$, these reduce to the Legendre functions discussed in Section 1.1.

Integral representations for the associated Legendre functions of the first and second kind are given by (cf. Erdélyi et al. 1953, p. 155 and p. 156, Nikolaev, 1970a)

$$P_{-\frac{1}{2}+i\tau}^m(x) = \frac{(-1)^m \sqrt{2}}{\pi \sqrt{\pi}} \Gamma(m + \frac{1}{2})(x^2 - 1)^{\frac{m}{2}} \cosh \pi \tau \int_0^\infty \frac{\cos \tau t \, dt}{(x + \cosh t)^{m+\frac{1}{2}}}, \tag{1.2.2}$$

8

$$Q_{-\frac{1}{2}+i\tau}^{m}(x) = \frac{(-1)^m}{\sqrt{2\pi}}(x^2-1)^{\frac{m}{2}}\Gamma(m+\frac{1}{2})\times$$

$$\times\left[\int_0^\pi \frac{\cosh\tau t\ dt}{(x-\cos t)^{m+\frac{1}{2}}} - i\sinh\pi\tau\int_0^\infty \frac{e^{-i\tau t}dt}{(x+\cosh t)^{m+\frac{1}{2}}}\right], \tag{1.2.3}$$

where $m = 0,1,2,\dots$ and $x > 1$;

$$P_{-\frac{1}{2}+i\tau}^{\mu}(\cosh\alpha) = \frac{2}{\sqrt{\pi}}\frac{(2\sinh\alpha)^\mu}{\Gamma(\frac{1}{2}-\mu)}\int_0^\alpha \frac{\cos\tau t\ dt}{[2(\cosh\alpha-\cosh t)]^{\frac{1}{2}+\mu}}, \quad \alpha > 0,\ \mathrm{Re}\,\mu < \frac{1}{2},$$
$$\tag{1.2.4}$$

$$Q_{-\frac{1}{2}+i\tau}^{\mu}(\cosh\alpha) = \frac{e^{i\pi\mu}\sqrt{\pi}(2\sinh\alpha)^\mu}{\Gamma(\frac{1}{2}-\mu)}\int_\alpha^\infty \frac{e^{-i\tau t}\ dt}{[2(\cosh t-\cosh\alpha)]^{\frac{1}{2}+\mu}},$$

$$\alpha > 0,\ \mathrm{Re}\,\mu < \frac{1}{2},\ \mathrm{Re}(\mu+i\tau) > -\frac{1}{2}. \tag{1.2.5}$$

Using formula (1.2.5) and the relation (cf. Hobson, 1931, p 197)

$$\pi e^{i\pi\mu}\sinh\pi\tau P_{-\frac{1}{2}+i\tau}^{\mu}(\cosh\alpha) = \sinh\pi\tau\sin\pi\mu\left[Q_{-\frac{1}{2}-i\tau}^{\mu}(\cosh\alpha) + Q_{-\frac{1}{2}+i\tau}^{\mu}(\cosh\alpha)\right]$$

$$-i\cosh\pi\tau\cos\pi\mu[Q_{-\frac{1}{2}-i\tau}^{\mu}(\cosh\alpha) - Q_{-\frac{1}{2}+i\tau}^{\mu}(\cosh\alpha)], \tag{1.2.6}$$

the following integral representation can be obtained (cf. Nikolaev, 1970a)

$$\sqrt{\frac{\pi}{2}}\sinh\pi\tau\frac{\Gamma(\frac{1}{2}-\mu)}{\sinh^\mu\alpha}P_{-\frac{1}{2}+i\tau}^{\mu}(\cosh\alpha) = \cosh\pi\tau\cos\pi\mu\int_\alpha^\infty \frac{\sin\tau t\ dt}{(\cosh t-\cosh\alpha)^{\frac{1}{2}+\mu}}$$

$$+ \sinh\pi\tau\sin\pi\mu\int_\alpha^\infty \frac{\cos\tau t\ dt}{(\cosh t-\cosh\alpha)^{\frac{1}{2}+\mu}}. \tag{1.2.7}$$

From formula (1.2.7), another integral representation can be deduced as (cf. Nikolaev, 1970a)

$$P_{-\frac{1}{2}+i\tau}^{-\mu}(\cosh\alpha) = \sqrt{\frac{2}{\pi}}\frac{\cos\pi\mu}{\sinh\pi\tau}\frac{\Gamma(\frac{1}{2}+\mu)\sinh^\mu\alpha}{\Gamma(\frac{1}{2}+i\tau+\mu)\Gamma(\frac{1}{2}-i\tau+\mu)}\int_\alpha^\infty \frac{\sin\tau t\ dt}{(\cosh t-\cosh\alpha)^{\frac{1}{2}+\mu}},$$

$$\mathrm{Re}\,\mu < \frac{1}{2},\ \mathrm{Re}(\mu\pm i\tau) > -\frac{1}{2}. \tag{1.2.8}$$

Another integral representation for the associated Legendre function of the first kind (cf. Nikolaev, 1970b) is

$$P_{-\frac{1}{2}+i\tau}^{-\mu}(\cosh\alpha) = \sqrt{\frac{2}{\pi}}\frac{\Gamma(\frac{1}{2}+\mu)\sinh^\mu\alpha}{\Gamma(\frac{1}{2}+i\tau+\mu)\Gamma(\frac{1}{2}-i\tau+\mu)}\int_0^\infty \frac{\cos\tau t\ dt}{(\cosh\alpha+\cosh t)^{\frac{1}{2}+\mu}}. \tag{1.2.9}$$

Representations of associated Legendre functions in terms of hypergeometric functions are

$$P^{i\sigma}_{-\frac{1}{2}+i\tau}(x) = \left(\frac{1+x}{1-x}\right)^{\frac{i\sigma}{2}} \frac{1}{\Gamma(1-i\sigma)} F\left(\frac{1}{2}+i\tau, \frac{1}{2}-i\tau; 1-i\sigma; \frac{1-x}{2}\right), \quad -1 < x < 1,$$

$$(1.2.10)$$

$$P^{-\mu}_{-\frac{1}{2}+i\tau}(x) = \left(\frac{1+x}{1-x}\right)^{-\frac{\mu}{2}} \frac{1}{\Gamma(1+\mu)} F\left(\frac{1}{2}+i\tau, \frac{1}{2}-i\tau; 1+\mu; \frac{1-x}{2}\right), \quad -1 < x < 1,$$

$$(1.2.11)$$

$$P^{-\mu}_{\nu-\frac{1}{2}}(\cos\theta) = \frac{\tan^{\mu}\frac{\theta}{2}}{\Gamma(1+\mu)} F\left(\frac{1}{2}-\nu, \frac{1}{2}+\nu; 1+\mu; \sin^2\frac{\theta}{2}\right), \quad 0 < \theta < \pi, \quad (1.2.12)$$

$$P^{-\mu}_{\nu-\frac{1}{2}}(-\cos\theta) = \frac{\cot^{\mu}\frac{\theta}{2}}{\Gamma(1+\mu)} F\left(\frac{1}{2}-\nu, \frac{1}{2}+\nu; 1+\mu; \cos^2\frac{\theta}{2}\right), \quad 0 < \theta < \pi, \quad (1.2.13)$$

$$G^{\mu}_1(\sinh\alpha, \nu) \equiv \frac{\exp(-\frac{\pi\mu}{2})P^{i\mu}_{\nu-\frac{1}{2}}(i\sinh\alpha) - \exp(\frac{\pi\mu}{2})P^{i\mu}_{\nu-\frac{1}{2}}(-i\sinh\alpha)}{2i}$$

$$= \frac{\Gamma(\frac{1}{2}+\nu+i\mu)\sin[\frac{\pi}{2}(\nu-\frac{1}{2}+i\mu)]}{\Gamma(1+\nu)\sqrt{2\pi\cosh\alpha}} \left\{\frac{F\left(\frac{1}{2}+i\mu, \frac{1}{2}-i\mu; 1+\nu; \frac{e^{\alpha}}{2\cosh\alpha}\right)}{\exp(-\alpha\nu)}\right.$$

$$\left. - \frac{F\left(\frac{1}{2}+i\mu, \frac{1}{2}-i\mu; 1+\nu; \frac{e^{-\alpha}}{2\cosh\alpha}\right)}{\exp(\alpha\nu)}\right\}, \quad \alpha \in [0, \infty), \quad (1.2.14)$$

$$G^{\mu}_2(\sinh\alpha, \nu) \equiv \frac{\exp(-\frac{\pi\mu}{2})P^{i\mu}_{\nu-\frac{1}{2}}(i\sinh\alpha) + \exp(\frac{\pi\mu}{2})P^{i\mu}_{\nu-\frac{1}{2}}(-i\sinh\alpha)}{2}$$

$$= \frac{\Gamma(\frac{1}{2}+\nu+i\mu)\cos[\frac{\pi}{2}(\nu-\frac{1}{2}+i\mu)]}{\Gamma(1+\nu)\sqrt{2\pi\cosh\alpha}} \left\{\frac{F\left(\frac{1}{2}+i\mu, \frac{1}{2}-i\mu; 1+\nu; \frac{e^{\alpha}}{2\cosh\alpha}\right)}{\exp(-\alpha\nu)}\right.$$

$$\left. + \frac{F\left(\frac{1}{2}+i\mu, \frac{1}{2}-i\mu; 1+\nu; \frac{e^{-\alpha}}{2\cosh\alpha}\right)}{\exp(\alpha\nu)}\right\}, \quad \alpha \in [0, \infty), \quad (1.2.15)$$

and

$$Q^{i\mu}_{\nu-\frac{1}{2}}(i\sinh\alpha) = \sqrt{\frac{\pi}{2\cosh\alpha}} \frac{\Gamma(\frac{1}{2}+\nu+i\mu)}{\Gamma(1+\nu)} \exp\left\{-\left[\frac{i\pi}{2}(\nu+\frac{1}{2}) + \pi\mu + \alpha\nu\right]\right\} \times$$

$$\times F\left(\frac{1}{2}+i\mu, \frac{1}{2}-i\mu; 1+\nu; \frac{e^{-\alpha}}{2\cosh\alpha}\right), \quad \alpha \in [0, \infty), \quad \nu \neq -\frac{1}{2}-i\mu-k, \quad (k = 0, 1, 2, \ldots),$$

$$(1.2.16)$$

where $F(a, b; c; z)$ is the hypergeometric function defined in (1.1.4).

The following two identities (cf. Nikolaev, 1970b) can easily be verified by using the well-known representation (1.2.9):

$$P_{-\frac{1}{2}+i\tau}^{-\mu}(y) = \frac{\Gamma(1+2\mu)}{\Gamma(\frac{1}{2}+i\tau+\mu)\Gamma(\frac{1}{2}-i\tau+\mu)} \times$$

$$\times \int_1^\infty \frac{(x^2-1)^{\frac{\mu}{2}}(y^2-1)^{\frac{\mu}{2}}}{(x+y)^{1+2\mu}} \, P_{-\frac{1}{2}+i\tau}^{-\mu}(x) \, dx, \quad \text{Re } \mu > -\frac{1}{2}, \ y \geq 0, \qquad (1.2.17)$$

and

$$P_{-\frac{1}{2}+i\tau}^{-\mu}(y) = \frac{2\Gamma(\frac{1}{2}+\mu)}{\Gamma(\frac{1}{4}+\frac{1}{2}\mu+\frac{i\tau}{2})\Gamma(\frac{1}{4}+\frac{1}{2}\mu-\frac{i\tau}{2})} \int_1^\infty \frac{(x^2-1)^{\frac{\mu}{2}}(y^2-1)^{\frac{\mu}{2}}}{(x^2+y^2-1)^{\frac{1}{2}+\mu}} P_{-\frac{1}{2}+i\tau}^{-\mu}(x) dx,$$

$$\text{Re } \mu > -\frac{1}{2}, \ y \geq 0. \qquad (1.2.18)$$

A useful relation (cf. Belichenko, 1987a) is

$$P_{\nu-\frac{1}{2}}^{-\mu}(\cos\theta) = \frac{\pi}{\sin\pi\mu} \left[\frac{P_{\nu-\frac{1}{2}}^{\mu}(-\cos\theta)}{\Gamma(\frac{1}{2}+\nu+\mu)\,\Gamma(\frac{1}{2}-\nu+\mu)} - \frac{\cos\pi\nu}{\pi} P_{\nu-\frac{1}{2}}^{-\mu}(-\cos\theta) \right].$$

$$(1.2.19)$$

Estimates and asymptotic expansions of associated Legendre functions

From the integral representations (1.2.2) and (1.2.3), it follows that (cf. Belova and Ufliand, 1970)

$$\left| P_{-\frac{1}{2}+i\tau}^{m}(x) \right| \leq \cosh\pi\tau \frac{(x^2-1)^{\frac{m}{2}}\Gamma(m+\frac{1}{2})}{(x+1)^m \Gamma(\frac{1}{2})} \, P_{-\frac{1}{2}}(x)$$

$$\leq O(1)\cosh\pi\tau \, x^{-\frac{1}{2}}\ln x, \quad x > 1, \qquad (1.2.20)$$

and

$$\left| Q_{-\frac{1}{2}+i\tau}^{m}(x) \right| \leq \cosh\pi\tau \left| Q_{-\frac{1}{2}}^{m}(x) \right| + \frac{1}{2}\sinh\pi\tau \left| P_{-\frac{1}{2}}^{m}(x) \right|$$

$$\leq O(1)\cosh\pi\tau \, x^{-\frac{1}{2}}\ln x, \quad x > 1. \qquad (1.2.21)$$

From the representations (1.2.4) and (1.2.5), it is found that (cf. Nikolaev, 1970a)

$$\left| P_{-\frac{1}{2}+i\tau}^{-\mu}(\cosh\alpha) \right| \leq \frac{\Gamma\left(\frac{1}{2}+\sigma\right)}{\left| \Gamma\left(\frac{1}{2}+\mu\right) \right|} P_{-\frac{1}{2}}^{-\sigma}(\cosh\alpha), \quad \alpha \geq 0, \sigma > -\frac{1}{2}, |\tau| > 0, \qquad (1.2.22)$$

and

$$\left|Q_{-\frac{1}{2}+i\tau}^{-\mu}(\cosh\alpha)\right| \leq \frac{\Gamma\left(\frac{1}{2}+\sigma\right)}{\left|\Gamma\left(\frac{1}{2}+\mu\right)\right|}Q_{-\frac{1}{2}}^{-\sigma}(\cosh\alpha), \quad \alpha \geq 0, |\sigma| < \frac{1}{2}, |\tau| > 0, \qquad (1.2.23)$$

where $\mu = \sigma + ip$, and σ and p are real numbers.

The relation (1.2.10) implies that $P_{-\frac{1}{2}+i\tau}^{i\sigma}(x)$ is continuous in the region defined by the inequalities $(-1 < x < 1, -\infty < \sigma < \infty)$ and satisfies the inequality

$$\left|P_{-\frac{1}{2}+i\tau}^{i\sigma}(x)\right| \leq \sqrt{\frac{\sinh \pi\sigma}{\pi\sigma}}P_{-\frac{1}{2}+i\tau}(x). \qquad (1.2.24)$$

The relations (1.2.12) and (1.2.13) show that $P_{\nu-\frac{1}{2}}^{-\mu}(\pm\cos\theta)$ is a continuous function of θ on $(0,\pi)$, and an entire function of the parameters μ and ν. It follows from the Mehler−Dirichlet representation (cf. Erdélyi et al. 1953, p. 160) that, for $0 < \theta < \pi$ and $\nu = s + i\tau$ with an arbitrary real part,

$$\left|P_{\nu-\frac{1}{2}}^{-i\sigma}(\pm\cos\theta)\right| \leq \sqrt{\cosh \pi\sigma}\, e^{|\tau|(\pi-\theta)}P_{-\frac{1}{2}}(\pm\cos\theta), \qquad (1.2.25)$$

where $\sigma \in (-T,T)$, T being an arbitrary positive real number. Using the properties of $P_{-\frac{1}{2}}(-\cos\theta)$, it is found that (cf. Belichenko, 1987a)

$$P_{-\frac{1}{2}}(-\cos\theta) = \begin{cases} O(1)\ln\sin\frac{\theta}{2}, & 0 < \theta < a, \quad (a > 0), \\ O(1), & a < \theta < \pi. \end{cases} \qquad (1.2.26)$$

Using the following estimate for the hypergeometric series (cf. Moshinskii, 1990)

$$\left|F\left(\frac{1}{2}+i\mu, \frac{1}{2}-i\mu; 1+i\tau; x\right)\right| = O(1)\, F\left(\frac{1}{2}, \frac{1}{2}; 1; x\right)$$

$$= O(1)\, K(\sqrt{x}), \quad x \in [0,1), \quad \tau \in [0,\infty),$$

where $K(z)$ is the complete elliptic integral of the first kind, and using the asymptotic properties of $K(z)$ and the gamma functions, from the representations (1.2.14) and (1.2.15), we have (cf. Moshinskii, 1990)

$$|G_n^{\mu}(x,\tau)| = \left(\frac{\sinh \pi\tau}{\pi\tau}\right)^{\frac{1}{2}}\begin{cases} O(1), & x \in (0,a), \quad (a > 0), \\ O(1)x^{-\frac{1}{2}}\ln(1+x), & x \in (a,\infty), \end{cases} \qquad (1.2.27)$$

where μ is real or complex, τ is a real number and $n=1,2$.

Introducing the following temporary notation

$$\phi_\nu^m(x) \equiv \frac{1}{2}\left[e^{\mp\frac{1}{2}i\pi m}\, P_{\nu-\frac{1}{2}}^{-m}(ix) + e^{\pm\frac{1}{2}i\pi m}\, P_{\nu-\frac{1}{2}}^{-m}(-ix)\right], \quad (x > \text{ or } < 0), \qquad (1.2.28)$$

and

$$\omega_\nu^m(x) \equiv (-1)^m \frac{1}{2}\left[e^{\frac{1}{2}i\pi m}\, Q_{\nu-\frac{1}{2}}^{-m}(ix) + e^{-\frac{1}{2}i\pi m}\, Q_{\nu-\frac{1}{2}}^{-m}(-ix)\right], \qquad (1.2.29)$$

from the definitions of the associated Legendre functions of the first and second kind, the representations of these in terms of hypergeometric functions are (cf. Lebedev and Skal'skaya, 1968)

$$\phi_\nu^m(x) = \frac{\sqrt{\pi}(x^2+1)^{-\frac{m}{2}}}{2^m\Gamma(\frac{3}{4}+\frac{m}{2}+\frac{\nu}{2})\Gamma(\frac{3}{4}+\frac{m}{2}-\frac{\nu}{2})}F\left(\frac{1}{4}-\frac{m}{2}+\frac{\nu}{2},\frac{1}{4}-\frac{m}{2}-\frac{\nu}{2};\frac{1}{2};-x^2\right),$$
$$(1.2.30)$$

$$\phi_\nu^m(x) = \frac{\Gamma(\frac{3}{4}-\frac{m}{2}+\frac{\nu}{2})(x^2+1)^{-\frac{1}{4}}}{2^{m-\nu+1}\Gamma(1+\nu)\Gamma(\frac{3}{4}+\frac{m}{2}-\frac{\nu}{2})} \times$$

$$\times \left\{(\sqrt{x^2+1}+x)^\nu\, F\left(\frac{1}{2}+m,\frac{1}{2}-m;1+\nu;\frac{\sqrt{x^2+1}+x}{2\sqrt{x^2+1}}\right)\right.$$

$$\left. + (\sqrt{x^2+1}-x)^\nu\, F\left(\frac{1}{2}+m,\frac{1}{2}-m;1+\nu;\frac{\sqrt{x^2+1}-x}{2\sqrt{x^2+1}}\right)\right\}, \qquad (1.2.31)$$

$$\omega_\nu^m(x) = \frac{\pi\Gamma(\frac{1}{4}-\frac{m}{2}+\frac{\nu}{2})(x^2+1)^{-\frac{\nu}{2}-\frac{1}{4}}}{2^{m+1}\Gamma(1+\nu)\Gamma(\frac{1}{4}+\frac{m}{2}-\frac{\nu}{2})}\, F\left(\frac{1}{4}+\frac{m}{2}+\frac{\nu}{2},\frac{1}{4}-\frac{m}{2}+\frac{\nu}{2};1+\nu;\frac{1}{x^2+1}\right),$$
$$(1.2.32)$$

and

$$\omega_\nu^m(x) = \frac{\pi\Gamma(\frac{1}{4}-\frac{m}{2}+\frac{\nu}{2})(x^2+1)^{-\frac{1}{4}}}{2^{m-\nu+1}\Gamma(1+\nu)\Gamma(\frac{1}{4}+\frac{m}{2}-\frac{\nu}{2})}(\sqrt{x^2+1}+x)^{-\nu} \times$$

$$\times F\left(\frac{1}{2}+m,\frac{1}{2}-m;1+\nu;\frac{\sqrt{x^2+1}-x}{2\sqrt{x^2+1}}\right). \qquad (1.2.33)$$

The representation (1.2.30) shows that the function $\phi_\nu^m(x)$ is continuous in x over the interval $(0,\infty)$ and that it is an entire function of the parameter ν. From the relation (1.2.31) we obtain the estimate (cf. Lebedev and Skal'skaya, 1968) as

$$|\phi_{i\tau}^m(x)| \leq O(1)(x^2+1)^{-\frac{1}{4}}\left\{F\left(\frac{1}{2},\frac{1}{2};1;\frac{\sqrt{x^2+1}+x}{2\sqrt{x^2+1}}\right)+F\left(\frac{1}{2},\frac{1}{2};1;\frac{\sqrt{x^2+1}-x}{2\sqrt{x^2+1}}\right)\right\}$$

$$=\begin{cases} O(1), \quad 0 \leq x \leq a, \\ O(1)x^{-\frac{1}{2}}\ln(1+x), \quad a \leq x < \infty, \ (a>0). \end{cases} \tag{1.2.34}$$

As $|\nu| \to \infty$ with $|\arg \nu| \leq \frac{\pi}{2}$, the asymptotic expansions of the associated Legendre functions presented above are (cf. Belova and Ufliand, 1970)

$$P_{\nu-\frac{1}{2}}^m(\cosh\alpha) = \frac{\nu^{m-\frac{1}{2}}}{\sqrt{2\pi\sinh\alpha}}\left\{\frac{[1+\sqrt{\cosh\alpha}\,O(|\nu|^{-1})]}{e^{-\nu\alpha}} \pm \frac{(-1)^m\,i\,[1+O(|\nu|^{-1})]}{e^{\nu\alpha}}\right\},$$
$$\tag{1.2.35}$$

$$Q_{\nu-\frac{1}{2}}^m(\cosh\alpha) = (-1)^m\,\nu^{m-\frac{1}{2}}\left(\frac{\pi}{2\sinh\alpha}\right)^{\frac{1}{2}}e^{-\nu\alpha}[1+O(|\nu|^{-1})], \tag{1.2.36}$$

and (cf. Lebedev and Skal'skaya, 1968)

$$\frac{2^{m-\nu}\,\Gamma(1+\nu)\,\Gamma(\frac{3}{4}+\frac{1}{2}m-\frac{1}{2}\nu)}{\Gamma(\frac{3}{4}-\frac{1}{2}m+\frac{1}{2}\nu)}\phi_\nu^m(\sinh\alpha) = \frac{1}{2\sqrt{\cosh\alpha}}\times$$

$$\times\left\{e^{\nu\alpha}[1+e^{2\alpha}O(|\nu|^{-1})]+e^{-\alpha\nu}[1+e^{-2\alpha}O(|\nu|^{-1})]\right\}, \quad \alpha \geq 0, \tag{1.2.37}$$

$$\frac{2^{m-\nu}\,\Gamma(1+\nu)\,\Gamma(\frac{1}{4}+\frac{1}{2}m-\frac{1}{2}\nu)}{\pi\Gamma(\frac{1}{4}-\frac{1}{2}m+\frac{1}{2}\nu)}\,\omega_\nu^m(\sinh\alpha) = \frac{e^{-\alpha\nu}}{2\sqrt{\cosh\alpha}}[1+e^{-2\alpha}O(|\nu|^{-1})], \quad \alpha \geq 0.$$
$$\tag{1.2.38}$$

Substituting the expression

$$\left[\frac{\sinh\alpha}{2(\cosh\alpha-\cosh t)}\right]^{\frac{1}{2}+\mu} = \left(\frac{\alpha}{\alpha^2-t^2}\right)^{\frac{1}{2}+\mu}\sum_{n=0}^{\infty}B_{n\mu}(\alpha)(\alpha^2-t^2)^n, \tag{1.2.39}$$

$$\left(B_{0\mu}(\alpha)=1, B_{1\mu}(\alpha)=(\frac{1}{2}+\mu)\frac{\coth\alpha-\frac{1}{\alpha}}{4\alpha},\ldots\right)$$

into equation (1.2.4) and integrating termwise, using the result

$$J_\nu(\alpha\tau) = \frac{2^{1-\nu}}{\sqrt{\pi}\Gamma(\frac{1}{2}+\nu)}\left(\frac{\tau}{\alpha}\right)^\nu\int_0^\alpha(\alpha^2-t^2)^{\nu-\frac{1}{2}}\cos\tau t\,dt, \text{ Re }\nu > -\frac{1}{2}, \tag{1.2.40}$$

where $J_\nu(z)$ represents the Bessel function of the first kind, we find (cf. Nikolaev, 1970a)

14

$$P_{-\frac{1}{2}+i\tau}^{-\mu}(\cosh\alpha) = \sqrt{\frac{\alpha}{\sinh\alpha}}\sum_{n=0}^{\infty}C_{n\mu}(\alpha)\frac{J_{n+\mu}(\alpha\tau)}{\tau^{n+\mu}}, \qquad (1.2.41)$$

where $C_{0\mu}(\alpha) = 1, C_{1\mu}(\alpha) = (\frac{1}{4}-\mu^2)\dfrac{\coth\alpha-\frac{1}{\alpha}}{2},\dots$

For large τ, the relation (1.2.41) may be considered as an asymptotic expansion which is valid for all $\alpha \geq 0$. The corresponding asymptotic expansion for associated Legendre functions of the second kind have the form

$$\left.\begin{aligned}Q_{-\frac{1}{2}+i\tau}^{-\mu}(\cosh\alpha) &= -\frac{i\pi}{2}\sqrt{\frac{\alpha}{\sinh\alpha}}\sum_{n=0}^{\infty}C_{n\mu}(\alpha)\frac{H_{n+\mu}^{(2)}(\alpha\tau)}{\tau^{n+\mu}}, \\[2mm] Q_{-\frac{1}{2}-i\tau}^{-\mu}(\cosh\alpha) &= \frac{i\pi}{2}\sqrt{\frac{\alpha}{\sinh\alpha}}\sum_{n=0}^{\infty}C_{n\mu}(\alpha)\frac{H_{n+\mu}^{(1)}(\alpha\tau)}{\tau^{n+\mu}},\end{aligned}\right\} \qquad (1.2.42)$$

where $H_{\mu}^{(n)}(z)$ $(n=1,2)$ are the Hankel functions and $C_{n\mu}(\alpha)$ have the same values as in the relation (1.2.41). Equations in (1.2.42) are obtained from the representation (1.2.5) by a similar technique, using the results

$$\left.\begin{aligned}H_{\nu}^{(1)}(\alpha\tau) &= \frac{2^{1+\nu}}{i\sqrt{\pi}}\left(\frac{\alpha}{\tau}\right)^{\nu}\frac{1}{\Gamma(\frac{1}{2}-\nu)}\int_{\alpha}^{\infty}\frac{e^{i\tau t}dt}{(t^2-\alpha^2)^{\frac{1}{2}+\nu}}, \\[2mm] H_{\nu}^{(2)}(\alpha\tau) &= -\frac{2^{1+\nu}}{i\sqrt{\pi}}\left(\frac{\alpha}{\tau}\right)^{\nu}\frac{1}{\Gamma(\frac{1}{2}-\nu)}\int_{\alpha}^{\infty}\frac{e^{-i\tau t}dt}{(t^2-\alpha^2)^{\frac{1}{2}+\nu}},\end{aligned}\right\} \qquad (1.2.43)$$

where $|\text{Re }\nu| < \frac{1}{2}$ and also the relations

$$H_{-\nu}^{(1)}(z) = e^{i\pi\nu}H_{\nu}^{(1)}(z),$$

$$H_{-\nu}^{(2)}(z) = e^{-i\pi\nu}H_{\nu}^{(2)}(z).$$

As $|\mu| \to \infty$, by virtue of the definition (1.2.11), the following asymptotic expansions (cf. Lebedev and Skal'skaya, 1986) are obtained as

$$\left.\begin{aligned}P_{-\frac{1}{2}+i\tau}^{-\mu}(x) &= \left(\frac{1+x}{1-x}\right)^{-\frac{\mu}{2}}\frac{1}{\Gamma(1+\mu)}[1+O(|\mu|^{-1})], \\[2mm] P_{-\frac{1}{2}+i\tau}^{-\mu}(-x) &= \left(\frac{1-x}{1+x}\right)^{-\frac{\mu}{2}}\frac{1}{\Gamma(1+\mu)}[1+O(|\mu|^{-1})],\end{aligned}\right\} \qquad (1.2.44)$$

15

where $z \in (-1, 1)$ and r is a real parameter. Also as $|\mu| \to \infty$,

$$
\left.
\begin{aligned}
P_{\nu-\frac{1}{2}}^{-\mu}(\cos\theta) &= \frac{\tan^{\mu}\frac{\theta}{2}}{\Gamma(1+\mu)}[1 + O(|\mu|^{-1})], \\[2ex]
P_{\nu-\frac{1}{2}}^{-\mu}(-\cos\theta) &= \frac{\cot^{\mu}\frac{\theta}{2}}{\Gamma(1+\mu)}[1 + O(|\mu|^{-1})],
\end{aligned}
\right\}
\tag{1.2.45}
$$

where $0 < \theta < \pi$, and

$$
P_{\nu}^{-\mu}(\cosh\alpha) = \frac{\tanh^{\mu}\frac{\alpha}{2}}{\Gamma(1+\mu)}[1 + O(|\mu|^{-1})],
\tag{1.2.46}
$$

$$
Q_{\nu}^{\mu}(\cosh\alpha) = \frac{1}{2}e^{i\pi\mu}\Gamma(1+\nu+\mu)\left[\frac{e^{\mp i\pi(1+\nu)}}{(1+\nu-\mu)^{1+\nu}} + \frac{e^{(1+\nu-\mu)\operatorname{sech}^2\frac{\alpha}{2}}}{(1+\nu-\mu)^{1+\nu}}\right] \times
$$

$$
\times \coth^{\mu}\frac{\alpha}{2}\,[1 + O(|\mu|^{-1})],
\tag{1.2.47}
$$

where the upper and lower sign is to be taken according as $-\frac{3\pi}{2} < \arg(1+\nu-\mu) < \frac{\pi}{2}$ or $-\frac{\pi}{2} < \arg(1+\nu-\mu) < \frac{3\pi}{2}$ and $\nu = -\frac{1}{2}+ir$ with $\operatorname{Re}\mu \geq 0$ (cf. Mandal, 1971b). The estimate (1.2.46) is obtained from a relation given in §3.2(3), p. 122 of Erdélyi et al. (1953) and the estimate (1.2.47) can be obtained from a relation given in §3.2(36), p. 132 of Erdélyi et al. (1953) along with the asymptotic expansion of $F(a,b;c;z)$ for large $|b|$ given on p. 77, formulae (14) and (15) of Erdélyi et al. (1953).

1.3 Some properties of generalized associated Legendre functions

The functions $P_k^{\mu,\nu}(z)$ and $Q_k^{\mu,\nu}(z)$, two specified linearly independent solutions of the differential equation

$$
(1-z^2)\frac{d^2u}{dz^2} - 2z\frac{du}{dz} + \left\{k(k+1) - \frac{\mu^2}{2(1-z)} - \frac{\nu^2}{2(1+z)}\right\}u = 0,
\tag{1.3.1}
$$

have been introduced by Kuipers and Meulenbeld (1957) as functions of z for all points of the complex z-plane, cut along the real axis from $-\infty$ to 1, the parameters k, μ and ν being in general complex. For $\mu = \nu$, the differential equation (1.3.1) reduces to the differential equation (1.2.1) whose solutions are associated Legendre

16

functions and for $\mu = \nu = 0$, equation (1.3.1) reduces to the differential equation (1.1.1) whose solutions are Legendre functions.

In terms of hypergeometric functions, the generalized associated Legendre functions are defined by (cf. Braaksma and Meulenbeld, 1967)

$$P_k^{\mu,\nu}(z) = (z-1)^{-\frac{1}{2}\mu}(z+1)^{\frac{1}{2}\nu}\frac{1}{\Gamma(1-\mu)}F\left(k+\frac{\nu-\mu}{2}+1,-k+\frac{\nu-\mu}{2};1-\mu;\frac{1-z}{2}\right),$$

(1.3.2)

for z not lying on the cut,

$$P_k^{\mu,\nu}(x) = \frac{(1+x)^{\frac{\nu}{2}}}{(1-x)^{\frac{\mu}{2}}}\frac{1}{\Gamma(1-\mu)}F\left(k+\frac{\nu-\mu}{2}+1,-k+\frac{\nu-\mu}{2};1-\mu;\frac{1-x}{2}\right), \quad (1.3.3)$$

$$P_k^{\mu,\nu}(x) = \frac{2^{-k+\frac{\nu-\mu}{2}}}{\Gamma(1-\mu)}\frac{(1+x)^{k+\frac{\mu}{2}}}{(1-x)^{\frac{\mu}{2}}}F\left(-k+\frac{\nu-\mu}{2},-k-\frac{\nu+\mu}{2};1-\mu;\frac{x-1}{x+1}\right), \quad (1.3.4)$$

for $-1 < x < 1$, and

$$Q_k^{\mu,\nu}(z) = e^{i\pi\mu}\,2^{k-\frac{1}{2}(\mu-\nu)}\,\frac{\Gamma(k+\frac{\nu+\mu}{2}+1)\,\Gamma(k+\frac{\mu-\nu}{2}+1)}{\Gamma(2k+2)}\,(z-1)^{-k-\frac{\mu}{2}-1}\times$$

$$\times(z+1)^{\frac{\nu}{2}}\,F\left(k+\frac{\mu+\nu}{2}+1,k-\frac{\mu-\nu}{2}+1;2k+2;\frac{2}{1-z}\right), \quad (1.3.5)$$

for z not lying on the cut.

Integral representations of the generalized associated Legendre functions of the first kind are given by (cf. Virchenko, 1984)

$$P_{-\frac{1}{2}+i\tau}^{\mu,\nu}(\cosh\alpha) = \frac{2^{\frac{\nu-\mu+1}{2}}\sinh^{\mu}\alpha}{\sqrt{\pi}\,\Gamma(\frac{1}{2}-\mu)}\int_0^{\alpha}(\cosh\alpha-\cosh\phi)^{-\mu-\frac{1}{2}}\times$$

$$\times F\left(\frac{\nu-\mu}{2},-\frac{\nu+\mu}{2};\frac{1}{2}-\mu;\frac{\cosh\alpha-\cosh\phi}{1+\cosh\alpha}\right)\cos\tau\phi\,d\phi, \quad (1.3.6)$$

and (cf. Mandal, 1995)

$$P_{-\frac{1}{2}+i\tau}^{\mu,\nu}(\cosh\alpha) = \frac{(2\pi)^{\frac{3}{2}}\,2^{\frac{\nu-\mu}{2}}\,\text{cosech}\,2\pi\tau\,\sinh^{-\mu}\alpha}{\Gamma(\frac{1}{2}+\mu)}\left\{\Gamma\left(\frac{1-\mu+\nu}{2}+i\tau\right)\times\right.$$

$$\left.\times\Gamma\left(\frac{1-\mu+\nu}{2}-i\tau\right)\Gamma\left(\frac{1-\mu-\nu}{2}+i\tau\right)\Gamma\left(\frac{1-\mu-\nu}{2}-i\tau\right)\right\}^{-1}\times$$

$$\times\int_{\alpha}^{\infty}(\cosh\phi-\cosh\alpha)^{\mu-\frac{1}{2}}F\left(\frac{\mu-\nu}{2},\frac{\mu+\nu}{2};\frac{1}{2}+\mu;\frac{\cosh\alpha-\cosh\phi}{1+\cosh\alpha}\right)\sin\tau\phi\,d\phi, \quad (1.3.7)$$

where $\operatorname{Re}\mu < \frac{1}{2}$, $|\operatorname{Re}\nu| < 1 - \operatorname{Re}\mu$.

The following relations exist between the generalized associated Legendre functions of the first and second kind (cf. Braaksma and Meulenbeld, 1967): For $x > -1$ and $x \neq 1$,

$$P_k^{\mu,\nu}(x) = 2^{k+\frac{\nu}{2}+1}\,(1+x)^{-\frac{1}{2}}\,P_{-\frac{1}{2}(\nu+1)}^{\mu,-2k-1}\left(\frac{x-3}{-x-1}\right). \tag{1.3.8}$$

For $-1 < x < 1$,

$$P_k^{\mu,\nu}(x) = \frac{2^{-k-\frac{\nu}{2}+1}}{\Gamma(-k-\frac{\mu+\nu}{2})\Gamma(k-\frac{\mu+\nu}{2}+1)}(1-x)^{-\frac{1}{2}}\,e^{i\pi\nu}\,Q_{-\frac{1}{2}(1+\mu)}^{-\nu,2k+1}\left(\frac{-x-3}{x-1}\right). \tag{1.3.9}$$

For $x > 1$,

$$Q_k^{\mu,\nu}(x) = \frac{\Gamma(k+\frac{\mu+\nu}{2}+1)\,\Gamma(k+\frac{\mu-\nu}{2}+1)}{2^{k-\frac{\nu}{2}+1}\,(1+x)^{\frac{1}{2}}\,e^{-i\pi\mu}}\,P_{\frac{1}{2}(\nu-1)}^{-2k-1,-\mu}\left(\frac{x-3}{x+1}\right), \tag{1.3.10}$$

$$P_k^{\mu,\nu}(x) = \frac{2^{k-\frac{\nu}{2}+2}\,(x-1)^{-\frac{1}{2}}\,e^{-i\pi(2k+1)}}{\Gamma(k-\frac{\mu-\nu}{2}+1)\Gamma(k-\frac{\mu+\nu}{2}+1)}\,Q_{-\frac{1}{2}(1+\mu)}^{2k+1,\nu}\left(\frac{x+3}{x-1}\right), \tag{1.3.11}$$

and

$$Q_k^{\mu,\nu}(x) = \frac{\Gamma(k+\frac{\mu+\nu}{2}+1)\,\Gamma(k+\frac{\mu-\nu}{2}+1)}{2^{k+\frac{\nu}{2}+1}\,(x-1)^{\frac{1}{2}}\,e^{-i\pi\mu}}\,P_{\frac{1}{2}(\mu-1)}^{-2k-1,\nu}\left(\frac{x+3}{x-1}\right). \tag{1.3.12}$$

Asymptotic expansions of $P_k^{\mu,\nu}(x)$ as $|\nu| \to \infty$ are given as follows (cf. Braaksma and Meulenbeld, 1967):

For fixed $\alpha > 0$, as $|\nu| \to \infty$

$$P_k^{\mu,\nu}(1 - 2\tanh^2\alpha) = \pi^{-\frac{1}{2}}2^{-\frac{1}{2}-\frac{3}{2}\mu+\frac{1}{2}\nu}(-\nu)^{\mu-\frac{1}{2}}(\tanh\alpha)^{-\frac{1}{2}}e^{-\nu\alpha}[1 + O(|\nu|^{-1})], \tag{1.3.13}$$

$$P_k^{\nu,\mu}(2\tanh^2\alpha - 1) = \frac{2^{\frac{\mu}{2}-\frac{3}{2}\nu}}{\Gamma(1-\nu)}(\tanh\alpha)^{-\frac{1}{2}}e^{\nu\alpha}\ [1 + O(|\nu|^{-1})], \tag{1.3.14}$$

where $|\arg(-\nu)| \leq \frac{\pi}{2} - \theta\ (0 < \theta < \frac{\pi}{2})$ and $|\arg(-\nu)| \leq \pi - \theta\ (0 < \theta < \frac{\pi}{2})$ respectively.

For fixed $\alpha > 0$, as $|\nu| \to \infty$ and $|\nu\alpha| \to \infty$

$$P_k^{\nu,\mu}(2\tanh^2\alpha - 1) = \frac{2^{\frac{\mu}{2}-\frac{3}{2}\nu}}{\Gamma(1-\nu)}(\tanh\alpha)^{-\frac{1}{2}}e^{\nu\alpha}[1 + (\nu\tanh\alpha)^{-1}O(1)$$

$$+(\tanh\alpha)^{-\mu-1}\,\nu^{-1}\,o(1) + (\tanh\alpha)^{\mu-1}\,\nu^{-1}\,o(1)], \tag{1.3.15}$$

18

$$P_k^{\mu,\nu}(1 - 2\tanh^2 \alpha) = \pi^{-\frac{1}{2}} 2^{-\frac{1}{2} - \frac{3}{2}\mu + \frac{1}{2}\nu}(-\nu)^{\mu - \frac{1}{2}}(\tanh \alpha)^{-\frac{1}{2}} \times$$

$$\times [e^{-\nu\alpha}\left\{1 + (\nu \tanh \alpha)^{-1}O(1) + (\tanh \alpha)^{-\mu - 1}\nu^{-1}o(1) + (\tanh \alpha)^{\mu - 1}\nu^{-1}o(1)\right\}$$

$$+ e^{\nu\alpha \pm i\pi(\mu - \frac{1}{2})}\left\{1 + (\nu \tanh \alpha)^{-1}O(1) + (\tanh \alpha)^{-\mu - 1}\nu^{-1}o(1) + (\tanh \alpha)^{\mu - 1}\nu^{-1}o(1)\right\}],$$

$$(1.3.16)$$

where $\theta \le |\arg \nu| \le \pi - \theta$. The upper or lower sign is to be taken according as Im $\nu >$ or < 0.

For $|\nu\alpha|$ bounded, as $|\nu| \to \infty$

$$P_k^{\nu,\mu}(2\tanh^2 \alpha - 1) = \frac{2^{-\frac{3}{2}\nu}\nu^{\frac{1}{2}}}{\Gamma(1 - \nu)}\left\{\alpha^\mu O(1) + \alpha^{-\mu}O(1)\right\}, \qquad (1.3.17)$$

$$P_k^{\mu,\nu}(1 - 2\tanh^2 \alpha) = 2^{\frac{k}{2}}(\nu)^\mu\left\{\alpha^\mu O(1) + \alpha^{-\mu}O(1)\right\}, \qquad (1.3.18)$$

in the entire complex ν-plane.

For $x > 1$, as $|k| \to \infty$ with $|\arg k| \le \pi - \theta$ $(0 < \theta < \pi)$

$$Q_k^{-\mu,-\nu}(x) = k^{-\mu - \frac{1}{2}}(x + \sqrt{x^2 - 1})^{-k}O(1). \qquad (1.3.19)$$

The above asymptotic formula (1.3.19) is obtained by using the formula

$$Q_k^{-\mu,-\nu}(x) = e^{-i\pi\mu}\, 2^{k + \frac{1}{2}(\mu - \nu)}\frac{\Gamma(k - \frac{\mu + \nu}{2} + 1)\,\Gamma(k - \frac{\mu - \nu}{2} + 1)}{\Gamma(2k + 2)} \times$$

$$\times (x + 1)^{-k - \frac{1}{2}\mu - 1}(x - 1)^{\frac{1}{2}\mu}F\left(k + \frac{\mu + \nu}{2} + 1, k + \frac{\mu - \nu}{2} + 1; 2k + 2; \frac{2}{1 + x}\right), \quad (1.3.20)$$

and the asymptotic expansion of the hypergeometric function

$$F\left(k + \frac{\mu + \nu}{2} + 1, k + \frac{\mu - \nu}{2} + 1; 2k + 2; \frac{2}{1 + x}\right) = 2^{2k}\left[1 + \left(\frac{x - 1}{x + 1}\right)^{\frac{1}{2}}\right]^{-2k} O(1),$$

$$(1.3.21)$$

as $|k| \to \infty$ with $|\arg k| \le \pi - \theta$, $(0 < \theta < \pi)$.

Some useful results

Let p and q be arbitrary complex numbers with $\operatorname{Re}(p - \frac{1}{2}\nu) > -1$ and $(\operatorname{Re} q + 1) > \frac{1}{2}|\operatorname{Re} \mu|$. Then we have (cf. Braaksma and Meulenbeld, 1967)

$$\int_{-1}^{1}(1-x)^p(1+x)^q \, P_k^{\nu,\mu}(x) \, dx = \frac{\Gamma(1+q+\frac{\mu}{2})\Gamma(1+p-\frac{1}{2}\nu)}{\Gamma(1-\nu)\Gamma(2+p+q+\frac{\mu-\nu}{2})} 2^{p+q+\frac{\mu-\nu}{2}+1} \times$$

$$\times {}_3F_2\left(k+\frac{\mu-\nu}{2}+1, -k+\frac{\mu-\nu}{2}, p-\frac{1}{2}\nu+1; 1-\nu, p+q+\frac{\mu-\nu}{2}+2; 1\right).$$

(1.3.22)

Choosing $q = k - p - 1$, relation (1.3.22) becomes

$$\int_{-1}^{1}(1-x)^p(1+x)^{k-p-1}P_k^{\nu,\mu}(x) \, dx = 2^{k+\frac{\mu-\nu}{2}} \times$$

$$\times \frac{\Gamma(k-p+\frac{1}{2}\mu) \, \Gamma(k-p-\frac{1}{2}\mu) \, \Gamma(p-\frac{1}{2}\nu+1)}{\Gamma(k+\frac{\mu-\nu}{2}+1) \, \Gamma(k-\frac{\mu+\nu}{2}+1) \, \Gamma(-p-\frac{1}{2}\nu)},$$

(1.3.23)

where $\mathrm{Re}(p-\frac{1}{2}\nu) > -1$ and $\mathrm{Re}(k-p) > \frac{1}{2}|\mathrm{Re}\,\mu|$.

Substituting $p = q = -\frac{1}{2}$, formula (1.3.22) reduces to

$$\int_{-1}^{1}\frac{P_k^{\nu,\mu}(x)}{(1-x^2)^{\frac{1}{2}}} \, dx = \frac{\pi^2 \, 2^{\frac{1}{2}\mu+\frac{1}{2}\nu}}{\cos\frac{\pi\mu}{2}}\left\{\Gamma\left(\frac{1}{2}-\frac{1}{2}k-\frac{1}{4}(\mu+\nu)\right) \Gamma\left(\frac{1}{2}-\frac{1}{2}k+\frac{1}{4}(\mu-\nu)\right) \times\right.$$

$$\left.\times \Gamma\left(1+\frac{1}{2}k+\frac{1}{4}(\mu-\nu)\right) \Gamma\left(1+\frac{1}{2}k-\frac{1}{4}(\mu+\nu)\right)\right\}^{-1},$$

(1.3.24)

where $\mathrm{Re}\,\nu < 1$, $|\mathrm{Re}\,\mu| < 1$.

Interchanging μ, ν and replacing x by $-x$, the relation (1.3.23) transforms to

$$\int_{-1}^{1}\frac{(1+x)^p P_k^{\mu,\nu}(-x)}{(1-x)^{p-k+1}} \, dx = 2^{k-\frac{\mu-\nu}{2}} \frac{\Gamma(k-p+\frac{1}{2}\nu)\Gamma(p-\frac{1}{2}\mu+1)}{\Gamma(1+k-\frac{\mu-\nu}{2})\Gamma(1+k-\frac{\mu+\nu}{2})} \frac{\Gamma(k-p-\frac{1}{2}\nu)}{\Gamma(-p-\frac{1}{2}\mu)},$$

(1.3.25)

where $\mathrm{Re}(p-\frac{1}{2}\mu) > -1$ and $\frac{1}{2}|\mathrm{Re}\,\nu| < \mathrm{Re}(k-p)$.

Chapter 2

Integral expansions related to Mehler—Fock transforms

In this chapter we present a number of integral expansions related to Mehler—Fock integral transforms for a class of functions involving spherical harmonics, namely, the Legendre, associated Legendre and generalized associated Legendre functions. In these expansion formulae, the subscript of the spherical harmonics appears as an integration variable in the inverse transformation formula while the superscript, whenever it exists, remains fixed. All these integral expansions may be found in the original literature. However, as they are not well documented, for the convenience of the reader these are presented in a unified manner in this chapter. The usefulness of these integral expansions is also demonstrated.

2.1 Integral expansions involving Legendre functions

In this section some integral expansions involving Legendre functions are presented.

2.1.1 The Mehler—Fock integral expansion of a function is provided by the following theorem (cf. Lebedev, 1965, p. 221).

Theorem 2.1.1

Let the function $f(x)$ be defined on $(1, \infty)$ and satisfy the conditions that

(i) $f(x)$ is piecewise continuous and has bounded variation in $(1, \infty)$,

(ii) the integrals $\int_1^a |f(x)| (x-1)^{-\frac{3}{4}} dx$ and $\int_a^\infty |f(x)| x^{-\frac{1}{2}} \ln x \, dx$,

are finite for every $a > 1$. Then at the points of continuity of $f(x)$,

$$ f(x) = \int_0^\infty \tau \tanh \pi\tau \ P_{-\frac{1}{2}+i\tau}(x) \left\{ \int_1^\infty f(y) \ P_{-\frac{1}{2}+i\tau}(y) \, dy \right\} d\tau. \qquad (2.1.1) $$

21

2.1.2 Lebedev and Skal'skaya (1966a) developed an integral expansion theorem, related to the above Mehler–Fock integral expansion for a class of functions, in terms of spherical functions. The expansion theorem is as follows.

Theorem 2.1.2

Let $f(x)$ be a function defined on $(-\infty, \infty)$ and let $f(x)$ satisfy the conditions that

(i) $f(x)$ is piecewise continuous and possesses a bounded variation in $(-\infty, \infty)$,

(ii) $f(x)|x|^{-\frac{1}{2}}\ln(1+|x|) \in L(-\infty, -a)$, $f(x)\,x^{-\frac{1}{2}}\ln(1+x) \in L(a, \infty)$, $(a > 0)$.

Then

$$\frac{1}{2}[f(x+0) + f(x-0)] = \int_0^\infty \frac{\tau\,\tanh \pi\tau}{\cosh \pi\tau} \left\{ \frac{P_{-\frac{1}{2}+i\tau}(ix) + P_{-\frac{1}{2}+i\tau}(-ix)}{2} \times \right.$$

$$\times \int_{-\infty}^\infty \frac{P_{-\frac{1}{2}+i\tau}(iy) + P_{-\frac{1}{2}+i\tau}(-iy)}{2} f(y)\,dy + \frac{P_{-\frac{1}{2}+i\tau}(ix) - P_{-\frac{1}{2}+i\tau}(-ix)}{2i} \times$$

$$\left. \times \int_{-\infty}^\infty \frac{P_{-\frac{1}{2}+i\tau}(iy) - P_{-\frac{1}{2}+i\tau}(-iy)}{2i} f(y)\,dy \right\} d\tau, \quad (-\infty < x < \infty). \tag{2.1.2}$$

In particular, for the even function

$$\frac{1}{2}[f(x+0) + f(x-0)] = 2\int_0^\infty \frac{\tau\,\tanh \pi\tau}{\cosh \pi\tau} \frac{P_{-\frac{1}{2}+i\tau}(ix) + P_{-\frac{1}{2}+i\tau}(-ix)}{2} \times$$

$$\times \left\{ \int_0^\infty \frac{P_{-\frac{1}{2}+i\tau}(iy) + P_{-\frac{1}{2}+i\tau}(-iy)}{2} f(y)\,dy \right\} d\tau, \quad (-\infty < x < \infty), \tag{2.1.3}$$

while for the odd function

$$\frac{1}{2}[f(x+0) + f(x-0)] = 2\int_0^\infty \frac{\tau\,\tanh \pi\tau}{\cosh \pi\tau} \frac{P_{\frac{1}{2}+i\tau}(ix) - P_{-\frac{1}{2}+i\tau}(-ix)}{2i} \times$$

$$\times \left\{ \int_0^\infty \frac{P_{-\frac{1}{2}+i\tau}(iy) - P_{-\frac{1}{2}+i\tau}(-iy)}{2i} f(y)\,dy \right\} d\tau, \quad (-\infty < x < \infty). \tag{2.1.4}$$

The last two formulae remain valid also for the functions defined on the interval $(0, \infty)$ and satisfying the following conditions:

(i) $f(x)$ is piecewise continuous and has bounded variation in the open interval $(0, \infty)$,

(ii) $f(x) \in L(0, a)$, $f(x)\,x^{-\frac{1}{2}}\ln(1+x) \in L(a, \infty)$, $(a > 0)$.

22

Proof:

The proof is given separately for the odd and the even functions. The validity of the theorem for an arbitary function $f(x)$ then automatically follows by writing $f(x)$ as a combination of even and odd functions, as given by

$$f(x) = \frac{1}{2}[f(x) + f(-x)] + \frac{1}{2}[f(x) - f(-x)].$$

Let us assume that $f(x)$ is an even function and let us consider the integral

$$J(x,T) = 2 \int_0^T \frac{\tau \, \tanh \pi\tau}{\cosh \pi\tau} \frac{P_{-\frac{1}{2}+i\tau}(ix) + P_{-\frac{1}{2}+i\tau}(-ix)}{2}$$

$$\times \left\{ \int_0^\infty \frac{P_{-\frac{1}{2}+i\tau}(iy) + P_{-\frac{1}{2}+i\tau}(-iy)}{2} f(y) \, dy \right\} d\tau, \quad (-\infty < x < \infty, \ T > 0). \quad (2.1.5)$$

From the estimates (1.1.10) and (1.1.19), it follows that

$$\int_0^\infty \left| \frac{P_{-\frac{1}{2}+i\tau}(iy) + P_{-\frac{1}{2}+i\tau}(-iy)}{2} f(y) \right| dy$$

$$\leq \left(\frac{\sinh \pi\tau}{\pi\tau} \right)^{\frac{1}{2}} \int_0^\infty \frac{P_{-\frac{1}{2}}(iy) + P_{-\frac{1}{2}}(-iy)}{2} |f(y)| \, dy$$

$$= O(1) \left\{ \int_0^a |f(y)| \, dy + \int_a^\infty y^{-\frac{1}{2}} \ln(1+y) |f(y)| \, dy \right\}.$$

This shows that under the conditions imposed on $f(x)$ as stated in the theorem, the integral converges uniformly. Hence, the integral under consideration is a continuous function of τ, and the repeated integral (2.1.5) is meaningful. Due to uniform convergence, one can change the order of integration and write $J(x,T)$ as

$$J(x,T) = \int_0^\infty K(x,y,T) \, f(y) \, dy, \quad (2.1.6)$$

where

$$K(x,y,T) = 2 \int_0^T \frac{\tau \, \tanh \pi\tau}{\cosh \pi\tau} \frac{\left\{ P_{-\frac{1}{2}+\tau}(ix) + P_{-\frac{1}{2}+i\tau}(-ix) \right\}}{2} \times$$

$$\times \frac{\left\{ P_{-\frac{1}{2}+i\tau}(iy) + P_{-\frac{1}{2}+i\tau}(-iy) \right\}}{2} d\tau. \quad (2.1.7)$$

Since the integrand is an even function of τ, we obtain

$$K(x,y,T) = i \int_{-iT}^{iT} \frac{\nu \, \tan \pi\nu}{\cos \pi\nu} \frac{\left\{ P_{\nu-\frac{1}{2}}(ix) + P_{\nu-\frac{1}{2}}(-ix) \right\} \left\{ P_{\nu-\frac{1}{2}}(iy) + P_{\nu-\frac{1}{2}}(-iy) \right\}}{4} d\nu.$$

$$(2.1.8)$$

23

Let x be a fixed positive number. Then using the relation

$$\pi \tan \pi\nu \; P_{\nu-\frac{1}{2}}(z) = Q_{-\nu-\frac{1}{2}}(z) - Q_{\nu-\frac{1}{2}}(z), \tag{2.1.9}$$

the kernel $K(x,y,T)$ can be represented in one of the following forms:

$$K(x,y,T) = \frac{1}{\pi i} \int_{-iT}^{iT} \frac{\nu\left\{P_{\nu-\frac{1}{2}}(iy) + P_{\nu-\frac{1}{2}}(-iy)\right\}\left\{Q_{\nu-\frac{1}{2}}(ix) + Q_{\nu-\frac{1}{2}}(-ix)\right\}}{2\cos \pi\nu} \, d\nu$$

$$(y \le x),$$

$$K(x,y,T) = \frac{1}{\pi i} \int_{-iT}^{iT} \frac{\nu\left\{P_{\nu-\frac{1}{2}}(ix) + P_{\nu-\frac{1}{2}}(-ix)\right\}\left\{Q_{\nu-\frac{1}{2}}(iy) + Q_{\nu-\frac{1}{2}}(-iy)\right\}}{2\cos \pi\nu} \, d\nu$$

$$(y \ge x). \tag{2.1.10}$$

In the integrands of (2.1.10), the zeros of the denominator $\nu = \frac{1}{2}(2m+1)$, $(m = 0,1,2,\ldots)$ cancel with the zeros of $Q_{\nu-\frac{1}{2}}(z) + Q_{\nu-\frac{1}{2}}(-z)$ when $m = 2n$ and with the zeros of $P_{\nu-\frac{1}{2}}(z) + P_{\nu-\frac{1}{2}}(-z)$ when $m = 2n+1$. Therefore, the integrands in (2.1.10) are analytic functions of ν, regular on the semi-plane Re $\nu \ge 0$. Hence, the integration along the segment of the imaginary axis can be replaced by integration along a semi-circle Γ_T of large radius T in the semi-plane Re $\nu \ge 0$.

Now we consider the behaviour of $J(x,T)$ as $T \to \infty$. Substituting $x = \sinh \alpha$, $y = \sinh \alpha'$, we can write $J(x,T)$ as

$$J(x,T) = \int_0^{\alpha} f(\sinh \alpha')K(\sinh \alpha, \sinh \alpha', T) \; \cosh \alpha' \, d\alpha'$$

$$+ \int_{\alpha}^{\infty} f(\sinh \alpha')K(\sinh \alpha, \sinh \alpha', T) \; \cosh \alpha' \, d\alpha'$$

$$= J_1(\alpha, T) + J_2(\alpha, T), \text{ say.} \tag{2.1.11}$$

Using the estimates (1.1.10), (1.1.12) and the asymptotic formulae for the gamma functions as $|\nu| \to \infty$ with $|\arg \nu| \le \frac{\pi}{2}$, we find that

$$\frac{\nu\left\{P_{\nu-\frac{1}{2}}(i\sinh \alpha') + P_{\nu-\frac{1}{2}}(-i\sinh \alpha')\right\}\left\{Q_{\nu-\frac{1}{2}}(i\sinh \alpha) + Q_{\nu-\frac{1}{2}}(-i\sinh \alpha)\right\}}{4\cos \pi\nu}$$

$$= \frac{\left\{e^{-(\alpha-\alpha')\nu} + e^{-(\alpha+\alpha')\nu} + e^{-(\alpha-\alpha')\nu} O(|\nu|^{-1}) + e^{-(\alpha+\alpha')\nu} O(|\nu|^{-1})\right\}}{4\sqrt{\cosh \alpha \cosh \alpha'}}, \quad (\alpha' \le \alpha).$$

$$\tag{2.1.12}$$

24

Assuming that $\nu = Te^{i\phi}$ $(-\frac{\pi}{2} \le \phi \le \frac{\pi}{2})$ holds along the arc Γ_T and using (2.1.12) and the known inequality

$$\int_0^{\frac{\pi}{2}} e^{-\lambda T \cos\phi}\, d\phi \le \frac{\pi}{2} \frac{1 - e^{-\lambda T}}{\lambda T} \quad (\lambda \ge 0), \tag{2.1.13}$$

we obtain that

$$K(\sinh\alpha, \sinh\alpha', T) = \frac{1}{\sqrt{\cosh\alpha\,\cosh\alpha'}} \left\{ \frac{\sin(\alpha-\alpha')T}{\pi(\alpha-\alpha')} + \frac{\sin(\alpha+\alpha')T}{\pi(\alpha+\alpha')} \right.$$

$$+ O(1)\frac{1-e^{-(\alpha-\alpha')T}}{(\alpha-\alpha')T} + O(1)\frac{1-e^{-(\alpha+\alpha')T}}{(\alpha+\alpha')T} \bigg\}, \quad \alpha' \le \alpha. \tag{2.1.14}$$

Using the estimate (2.1.14) in J_1, we find that

$$J_1(\alpha, T) = \frac{1}{\pi}\int_0^{\alpha} f(\sinh\alpha') \left(\frac{\cosh\alpha'}{\cosh\alpha}\right)^{\frac{1}{2}} \frac{\sin(\alpha-\alpha')T}{(\alpha-\alpha')}\, d\alpha'$$

$$+ \frac{1}{\pi}\int_0^{\alpha} f(\sinh\alpha') \left(\frac{\cosh\alpha'}{\cosh\alpha}\right)^{\frac{1}{2}} \frac{\sin(\alpha+\alpha')T}{(\alpha+\alpha')}\, d\alpha'$$

$$+ O(1)\int_0^{\alpha} |f(\sinh\alpha')| \left(\frac{\cosh\alpha'}{\cosh\alpha}\right)^{\frac{1}{2}} \frac{1-e^{-(\alpha-\alpha')T}}{(\alpha-\alpha')T}\, d\alpha'$$

$$+ O(1)\int_0^{\alpha} |f(\sinh\alpha')| \left(\frac{\cosh\alpha'}{\cosh\alpha}\right)^{\frac{1}{2}} \frac{1-e^{-(\alpha+\alpha')T}}{(\alpha+\alpha')T}\, d\alpha'. \tag{2.1.15}$$

The conditions prescribed for $f(x)$ imply that $f(\sinh\alpha')\sqrt{\cosh\alpha'} \in L(0,\infty)$ and then as $T \to \infty$, we obtain

$$\frac{1}{\pi}\int_0^{\alpha} f(\sinh\alpha') \left(\frac{\cosh\alpha'}{\cosh\alpha}\right)^{\frac{1}{2}} \frac{\sin(\alpha-\alpha')T}{(\alpha-\alpha')}\, d\alpha' = \frac{1}{2}f(\sinh\alpha - 0) + o(1), \tag{2.1.16}$$

$$\frac{1}{\pi}\int_0^{\alpha} f(\sinh\alpha') \left(\frac{\cosh\alpha'}{\cosh\alpha}\right)^{\frac{1}{2}} \frac{\sin(\alpha+\alpha')T}{(\alpha+\alpha')}\, d\alpha' = o(1). \tag{2.1.17}$$

Further, if the interval of integration is divided into the subintervals $(0, \alpha - \delta)$ and $(\alpha - \delta, \alpha)$ where δ is chosen to be sufficiently small and positive, and if we use the inequality

$$\frac{1-e^{-(\alpha-\alpha')T}}{(\alpha-\alpha')T} \le \begin{cases} \frac{1}{\delta T}, & 0 < \alpha' < \alpha - \delta, \\ 1, & \alpha - \delta < \alpha' < \alpha, \end{cases}$$

we find that

$$\int_0^\alpha |f(\sinh \alpha')| \left(\frac{\cosh \alpha'}{\cosh \alpha}\right)^{\frac{1}{2}} \frac{1 - e^{-(\alpha-\alpha')T}}{(\alpha - \alpha')T} \, d\alpha'$$

$$\leq \frac{1}{\delta T} \int_0^{\alpha-\delta} |f(\sinh \alpha')| \sqrt{\cosh \alpha'} \, d\alpha' + \int_{\alpha-\delta}^\alpha |f(\sinh \alpha')| \sqrt{\cosh \alpha'} \, d\alpha'$$

$$= O(T^{-1}) + o(1) = o(1) \quad \text{as} \quad T \to \infty, \tag{2.1.18}$$

and finally

$$\int_0^\alpha |f(\sinh \alpha')| \left(\frac{\cosh \alpha'}{\cosh \alpha}\right)^{\frac{1}{2}} \frac{1 - e^{-(\alpha+\alpha')T}}{(\alpha + \alpha')T} \, d\alpha'$$

$$\leq \frac{1}{\alpha T} \int_0^\alpha |f(\sinh \alpha')| \sqrt{\cosh \alpha'} \, d\alpha'$$

$$= O(T^{-1}) = o(1) \quad \text{as} \quad T \to \infty. \tag{2.1.19}$$

From (2.1.15) to (2.1.19), it follows that

$$\lim_{T \to \infty} J_1(\alpha, T) = \frac{1}{2} f(\sinh \alpha - 0). \tag{2.1.20}$$

In a similar manner, for J_2 we obtain

$$J_2(\alpha, T) = \frac{1}{\pi} \int_\alpha^\infty f(\sinh \alpha') \left(\frac{\cosh \alpha'}{\cosh \alpha}\right)^{\frac{1}{2}} \frac{\sin(\alpha' - \alpha)T}{(\alpha' - \alpha)} \, d\alpha'$$

$$+ \frac{1}{\pi} \int_\alpha^\infty f(\sinh \alpha') \left(\frac{\cosh \alpha'}{\cosh \alpha}\right)^{\frac{1}{2}} \frac{\sin(\alpha' + \alpha)T}{(\alpha' + \alpha)} \, d\alpha'$$

$$+ O(1) \int_\alpha^\infty |f(\sinh \alpha')| \left(\frac{\cosh \alpha'}{\cosh \alpha}\right)^{\frac{1}{2}} \frac{1 - e^{-(\alpha'-\alpha)T}}{(\alpha' - \alpha)T} \, d\alpha'$$

$$+ O(1) \int_\alpha^\infty |f(\sinh \alpha')| \left(\frac{\cosh \alpha'}{\cosh \alpha}\right)^{\frac{1}{2}} \frac{1 - e^{-(\alpha'+\alpha)T}}{(\alpha' + \alpha)T} \, d\alpha'. \tag{2.1.21}$$

Thus, it follows that

$$\lim_{T \to \infty} J_2(\alpha, T) = \frac{1}{2} f(\sinh \alpha + 0). \tag{2.1.22}$$

Therefore,

$$\lim_{T \to \infty} J(x, T) = \frac{1}{2}[f(x+0) + f(x-0)]. \tag{2.1.23}$$

The relation (2.1.23) is proved for positive x. Its validity for $x < 0$ follows from the fact that the functions $J(x, T)$ and $f(x)$ are even functions of x. The relation (2.1.23) can also be shown to be valid for $x = 0$ by a slight change in the above arguments.

Hence, we have proved the expansion Theorem 2.1.2 for even functions. For odd functions, the proof is analogous. In this case, the inequality (1.1.10) and the estimates (1.1.15) and (1.1.17), however, must be replaced by the inequality (1.1.11) and the estimates (1.1.16) and (1.1.18), respectively.

2.1.3   As examples of the above expansion formulae (2.1.2) – (2.1.4), we consider integral expansions of some simple functions (cf. Lebedev and Skal'skaya, 1966a).

Example 2.1.1

Let

$$f(x) = \begin{cases} 1, & |x| < a, \\ 0, & |x| > a, \ (a \text{ is any real number}). \end{cases}$$

Using the identity (1.1.20) and applying the integral expansion (2.1.3), we find

$$f(x) = -2\sqrt{a^2 + 1} \int_0^\infty \frac{\tau \tanh \pi\tau}{\left(\frac{1}{4} + \tau^2\right) \cosh \pi\tau} \cdot \frac{P^1_{-\frac{1}{2}+i\tau}(ia) + P^1_{-\frac{1}{2}+i\tau}(-ia)}{2} \times$$

$$\times \frac{P_{-\frac{1}{2}+i\tau}(ix) + P_{-\frac{1}{2}+i\tau}(-ix)}{2} \, d\tau, \quad -\infty < x < \infty. \tag{2.1.24}$$

Example 2.1.2

Utilizing the identity

$$\int_0^\infty \frac{P_{-\frac{1}{2}+i\tau}(ix) + P_{-\frac{1}{2}+i\tau}(-ix)}{2} \frac{dx}{\sqrt{x^2 + 1}} = \frac{\pi}{\cosh \pi\tau} \, [P_{-\frac{1}{2}+i\tau}(0)]^2, \tag{2.1.25}$$

we obtain

$$\frac{1}{\sqrt{x^2 + 1}} = 2\pi \int_0^\infty \frac{\tau \tanh \pi\tau}{\cosh^2 \pi\tau} \left\{ P_{-\frac{1}{2}+i\tau}(0) \right\}^2 \frac{[P_{-\frac{1}{2}+i\tau}(ix) + P_{-\frac{1}{2}+i\tau}(-ix)]}{2} \, d\tau,$$

$$(-\infty < x < \infty). \tag{2.1.26}$$

27

Example 2.1.3

The following expansion represents a generalization of (2.1.26):

$$\frac{1}{\sqrt{x^2 + a^2 + 1}} = 2\pi \int_0^\infty \frac{\tau \tanh \pi \tau}{\cosh^2 \pi \tau} P_{-\frac{1}{2}+i\tau}(0) \frac{[P_{-\frac{1}{2}+i\tau}(ia) + P_{-\frac{1}{2}+i\tau}(-ia)]}{2} \times$$

$$\times \frac{P_{-\frac{1}{2}+i\tau}(ix) + P_{-\frac{1}{2}+i\tau}(-ix)}{2} d\tau, \quad -\infty < x < \infty. \tag{2.1.27}$$

Example 2.1.4

Let

$$f(x) = \begin{cases} (a^2 - x^2)^{-\frac{1}{2}}, & |x| < a, \\ 0, & |x| > a. \end{cases}$$

Then

$$f(x) = \pi \int_0^\infty \frac{\tau \tanh \pi \tau}{\cosh \pi \tau} P_{-\frac{1}{2}+i\tau}(0) \, P_{-\frac{1}{2}+i\tau}(\sqrt{a^2 + 1}) \frac{[P_{-\frac{1}{2}+i\tau}(ix) + P_{-\frac{1}{2}+i\tau}(-ix)]}{2} d\tau.$$

$$\tag{2.1.28}$$

Although the assumption of piecewise continuity of the function is violated in this case at the points $x = \pm a$, the expansions (2.1.2) − (2.1.4) remain valid.

As applications of the expansion Theorem 2.1.2 discussed above, Lebedev and Skal'skaya (1966b) presented a technique of solving boudary value problems of potential and elasticity theory for hyperboloids of revolution of one sheet. This technique is based on the application of the integral expansion formula (2.1.2).

Example 2.1.5

Consider the Dirichlet problem for a hyperboloid of revolution of one sheet. Let (α, β, ϕ) denote oblate spheroidal coordinates connected with cylindrical coordinates (r, z, ϕ) by

$$r = c \cosh \alpha \sin \beta, \quad z = c \sinh \alpha \cos \beta, \quad (-\infty < \alpha < \infty, \quad 0 \le \beta \le \frac{\pi}{2}).$$

Laplace's equation in this coordinate system is

$$\frac{1}{\cosh \alpha} \frac{\partial}{\partial \alpha} \left(\cosh \alpha \frac{\partial u}{\partial \alpha} \right) + \frac{1}{\sin \beta} \frac{\partial}{\partial \beta} \left(\sin \beta \frac{\partial u}{\partial \beta} \right) + \left(\frac{1}{\cosh^2 \beta} - \frac{1}{\cosh^2 \alpha} \right) \frac{\partial^2 u}{\partial \phi^2} = 0$$

and its solution is chosen in the form

$$u = u_\tau(\alpha, \beta) = [A(\tau)\ P_{-\frac{1}{2}+i\tau}(i\sinh\alpha) + B(\tau)\ P_{-\frac{1}{2}+i\tau}(-i\sinh\alpha)]\times$$

$$\times[C(\tau)\ P_{-\frac{1}{2}+i\tau}(\cos\beta) + D(\tau)\ P_{-\frac{1}{2}+i\tau}(-\cos\beta)], \quad (0 \le \tau < \infty).$$

If we consider the inner Dirichlet problem, then the solutions must be bounded on the symmetry axis $r = 0$. The corresponding solution becomes

$$u = u_\tau(\alpha, \beta) = [M(\tau)P_{-\frac{1}{2}+i\tau}(i\sinh\alpha) + N(\tau)P_{-\frac{1}{2}+i\tau}(-i\sinh\alpha)]\ P_{-\frac{1}{2}+i\tau}(\cos\beta),$$

$$(-\infty < \alpha < \infty,\ 0 \le \beta < \frac{\pi}{2},\ 0 \le \tau < \infty).$$

In the case of the outer Dirichlet problem, ∇u must be bounded on the circle $r = c, z = 0$ and then the solution has the form

$$u = u_\tau(\alpha, \beta) = M(\tau)[P_{-\frac{1}{2}+i\tau}(i\sinh\alpha) + P_{-\frac{1}{2}+i\tau}(-i\sinh\alpha)]\times$$

$$[P_{-\frac{1}{2}+i\tau}(\cos\beta) + P_{-\frac{1}{2}+i\tau}(-\cos\beta)] + N(\tau)[P_{-\frac{1}{2}+i\tau}(i\sinh\alpha) - P_{-\frac{1}{2}+i\tau}(-i\sinh\alpha)]\times$$

$$\times[P_{-\frac{1}{2}+i\tau}(\cos\beta) - P_{-\frac{1}{2}+i\tau}(-\cos\beta)], (-\infty < \alpha < \infty,\ 0 < \beta \le \frac{\pi}{2}, 0 \le \tau < \infty).$$

Now the inner Dirichlet problem for a hyperboloid of revolution $\beta = \beta_0$ can be formulated as follows. We require to find a function u harmonic in $0 \le \beta \le \beta_0$, continuous in the closed region $0 \le \beta \le \beta_0$, satisfying the boundary conditions

$$u = f(\alpha) \text{ on } \beta = \beta_0, \tag{2.1.29}$$

$$u \to 0 \text{ as } |\alpha| \to \infty, \tag{2.1.30}$$

uniformly in β. We assume that the given function $f(\alpha)$ is continuous on $(-\infty, \infty)$ and tends to zero as $|\alpha| \to \infty$.

We restrict ourselves to the case when $f(\alpha)$ is an even function. The case when $f(\alpha)$ is an odd function of α can be dealt with in an analogous manner. A solution of the above boudary value problem can thus be expressed in the form

$$u = 2 \int_0^\infty \tau \, \tanh \pi\tau \, \frac{P_{-\frac{1}{2}+i\tau}(\cos\beta)}{P_{-\frac{1}{2}+i\tau}(\cos\beta_0)} \, \frac{[P_{-\frac{1}{2}+i\tau}(i\sinh\alpha) + P_{-\frac{1}{2}+i\tau}(-i\sinh\alpha)]}{2} \, F(\tau) \, d\tau,$$

$$(-\infty < \alpha < \infty, \quad 0 \le \beta < \beta_0), \qquad (2.1.31)$$

where $F(\tau)$ is to be determined.

Use of the boundary condition (2.1.29) in the representation (2.1.31) produces

$$f(\alpha) = 2 \int_0^\infty \tau \, \tanh \pi\tau \, \frac{[P_{-\frac{1}{2}+i\tau}(i\sinh\alpha) + P_{-\frac{1}{2}+i\tau}(-i\sinh\alpha)]}{2} \, F(\tau) \, d\tau,$$

$$(-\infty < \alpha < \infty), \qquad (2.1.32)$$

from which, using the expansion formula (2.1.3), we get

$$F(\tau) = \int_0^\infty f(\alpha) \, \frac{[P_{-\frac{1}{2}+i\tau}(i\sinh\alpha) + P_{-\frac{1}{2}+i\tau}(-i\sinh\alpha)]}{2} \, \cosh\alpha \, d\alpha. \qquad (2.1.33)$$

Equations (2.1.31) and (2.1.33) together give a formal solution to the problem. To prove this result rigorously, we assume that the given function $f(\alpha)$ can be represented by the relation (2.1.32) with $F(\tau)$ defined by (2.1.33) and

$$\sqrt{\tau} \, e^{-\frac{\pi\tau}{2}} F(\tau) \in L(0, \infty). \qquad (2.1.34)$$

Then, from the estimate

$$\frac{P_{-\frac{1}{2}+i\tau}(\cos\beta)}{P_{-\frac{1}{2}+i\tau}(\cos\beta_0)} \le 1, \quad (0 \le \beta \le \beta_0),$$

and the inequality (1.1.10), we obtain

$$2 \int_0^\infty \left| \tau \, \tanh \pi\tau \, \frac{P_{-\frac{1}{2}+i\tau}(\cos\beta)}{P_{-\frac{1}{2}+i\tau}(\cos\beta_0)} \, \frac{[P_{-\frac{1}{2}+i\tau}(i\sinh\alpha) + P_{-\frac{1}{2}+i\tau}(-i\sinh\alpha)]}{2} \, F(\tau) \right| d\tau$$

$$\le \frac{2\sqrt{2}}{\sqrt{\pi}} \frac{[P_{-\frac{1}{2}}(i\sinh\alpha) + P_{-\frac{1}{2}}(-i\sinh\alpha)]}{2} \int_0^\infty \sqrt{\tau} \, e^{-\frac{\pi\tau}{2}} |F(\tau)| \, d\tau.$$

This shows that the integral (2.1.31) converges absolutely. In the last estimate, the α-dependent factor is bounded for any α, hence convergence of the integral (2.1.31) is uniform on any closed region D $(-A \le \alpha \le A, \; 0 \le \beta \le \beta_0)$, where A is an arbitrary large constant. Since the integrand is harmonic in D, using Harnack's Theorem, we conclude that u is also harmonic within the hyperboloid $\beta = \beta_0$. Uniform convergence also implies the possibility of a limiting process $\beta \to \beta_0$ under the integral sign in (2.1.31). Therefore, by the relation (2.1.32), we find that

$$u(\alpha, \beta_0) = 2 \int_0^\infty \frac{\tau \tanh \pi \tau}{\cosh \pi \tau} \frac{[P_{-\frac{1}{2}+i\tau}(i \sinh \alpha) + P_{-\frac{1}{2}+i\tau}(-i \sinh \alpha)]}{2} F(\tau) \, d\tau = f(\alpha).$$

Finally, from (2.1.34) it follows that

$$u = O(1) \frac{[P_{-\frac{1}{2}}(i \sinh \alpha) + P_{-\frac{1}{2}}(-i \sinh \alpha)]}{2} \to 0 \text{ as } |\alpha| \to \infty,$$

uniformly in β. Thus, it is found that the expression for u defined by the integral (2.1.31) satisfies all the required conditions and is a solution of the aforesaid Dirichlet problem.

As another application of the expansion formula (2.1.2), Lebedev and Skal'skaya (1966b) considered an electrostatic problem on the distribution of charges induced on the surface of a hollow conducting hyperboloid and a problem of torsion of a shaft which has the form of a hyperboloid of revolution.

2.1.4 Moshinskii (1989) established some new expansion formulae generalizing those obtained by Lebedev and Skal'skaya (1966a) discussed above. He developed the following integral expansion theorem.

Theorem 2.1.3

Let $f(x)$ be defined on (ξ, ∞) $(-\infty < \xi < \infty)$ and satisfy the following conditions:

(i) $f(x)$ is piecewise continuous and has bounded variation in (ξ, b), $b < \infty$,

(ii) $f(x) \in L(\xi, a)$ and $f(x) x^{-\frac{1}{2}} \ln x \in L(a, \infty)$, $a > \xi$, $a > 0$.

Then, at the points of continuity of $f(x)$, we have

$$f(x) = \frac{i \pi^2}{4} \int_0^\infty \frac{\tau \sinh \pi \tau \, G_{i\tau}(x, \xi)}{\cosh^3 \pi \tau \, Q_{-\frac{1}{2}+i\tau}(i\xi) Q_{-\frac{1}{2}-i\tau}(i\xi)} \left\{ \int_\xi^\infty f(y) \, G_{i\tau}(y, \xi) \, dy \right\} d\tau,$$

$$(2.1.35)$$

31

where

$$G_\nu(x,y) = P_{\nu-\frac{1}{2}}(ix)\, P_{\nu-\frac{1}{2}}(-iy) - P_{\nu-\frac{1}{2}}(iy)\, P_{\nu-\frac{1}{2}}(-ix)$$

$$= \frac{2\cos\pi\nu}{\pi}\left[P_{\nu-\frac{1}{2}}(ix)\, Q_{\nu-\frac{1}{2}}(iy) - P_{\nu-\frac{1}{2}}(iy)\, Q_{\nu-\frac{1}{2}}(ix) \right]. \tag{2.1.36}$$

Proof:

The proof of this expansion theorem can be established in a manner similar to Theorem 2.1.2. To prove the theorem, we consider the following integral

$$J(x,T) = \frac{i\,\pi^2}{4}\int_0^T \frac{\tau\,\sinh\pi\tau\,\, G_{i\tau}(x,\xi)}{\cosh^3\pi\tau\,\, Q_{-\frac{1}{2}+i\tau}(i\xi)\, Q_{-\frac{1}{2}-i\tau}(i\xi)} \left\{ \int_\xi^\infty f(y)\, G_{i\tau}(y,\xi)\, dy \right\} d\tau. \tag{2.1.37}$$

Using the estimate (1.1.25), it follows that

$$\int_\xi^\infty |f(y)\, G_{i\tau}(y,\xi)|\, dy \le O(1)\,\frac{\sinh\pi\tau}{\pi\tau}\left\{ \int_\xi^a |f(y)|\, dy + \int_a^\infty |f(y)|\, y^{-\frac{1}{2}}\,\ln y\, dy \right\},$$

and this proves that under the conditions imposed on the function $f(x)$, the integral is majorized by a convergent integral. Hence, the integral represents a continuous function of τ, and so the integral $J(x,T)$ has a meaning. Changing the order of integration, we write

$$J(x,T) = \int_\xi^\infty K(x,y,T)\, f(y)\, dy, \tag{2.1.38}$$

where

$$K(x,y,T) = \frac{i\,\pi^2}{4}\int_0^T \frac{\tau\,\sinh\pi\tau\,\, G_{i\tau}(x,\xi)\, G_{i\tau}(y,\xi)}{\cosh^3\pi\tau\,\, Q_{-\frac{1}{2}+i\tau}(i\xi)\, Q_{-\frac{1}{2}-i\tau}(i\xi)}\, d\tau. \tag{2.1.39}$$

Now fix x ($x > \xi$). Since the integrand in (2.1.39) is an even function of τ, using the relation (2.1.9), we obtain

$$\left.\begin{aligned}
K(x,y,T) &= \frac{1}{2}\int_{-iT}^{iT} \frac{\nu\, Q_{\nu-\frac{1}{2}}(ix)\, G_\nu(y,\xi)}{\cos\pi\nu\, Q_{\nu-\frac{1}{2}}(i\xi)}\, d\nu, \quad x \ge y, \\[2mm]
K(x,y,T) &= \frac{1}{2}\int_{-iT}^{iT} \frac{\nu\, Q_{\nu-\frac{1}{2}}(iy)\, G_\nu(x,\xi)}{\cos\pi\nu\, Q_{\nu-\frac{1}{2}}(i\xi)}\, d\nu, \quad x \le y.
\end{aligned}\right\} \tag{2.1.40}$$

32

The integrands in (2.1.40) have no singularities in the half-plane Re $\nu \geq 0$, because the zeros of $\cos \pi \nu$ in this half-plane are cancelled by the zeros of $G_\nu(x, \xi)$, and $Q_{\nu-\frac{1}{2}}(ix)$ has neither zeros nor poles in this half-plane. As the integrand is regular in the half-plane Re $\nu \geq 0$, we replace the integration on the imaginary axis segment $(-iT, iT)$ by integration on the semicircle Γ_T of radius T in the half-plane Re $\nu \geq 0$. Now, we investigate the behaviour of the integral J as $T \to \infty$. Putting $x = \sinh \alpha$, $\xi = \sinh \beta$ and $y = \sinh \gamma$, and dividing the integral J into two parts, we have

$$
\begin{aligned}
J(x, T) &= \int_\beta^\infty f(\sinh \gamma)\, K(\sinh \alpha, \sinh \gamma, T)\, \cosh \gamma\, d\gamma \\
&= \int_\beta^\alpha f(\sinh \gamma)\, K(\sinh \alpha, \sinh \gamma, T)\, \cosh \gamma\, d\gamma \\
&\quad + \int_\alpha^\infty f(\sinh \gamma)\, K(\sinh \alpha, \sinh \gamma, T)\, \cosh \gamma\, d\gamma \\
&= J_1(\alpha, T) + J_2(\alpha, T), \quad \text{say.}
\end{aligned}
\tag{2.1.41}
$$

Following a similar analysis utilized in the proof of the expansion Theorem 2.1.2 and using the estimates (1.1.26), (1.1.27), (2.1.13) and asymptotic properties of the gamma functions, when $\nu = T\, \exp(i\phi)$, $\phi \in [-\frac{\pi}{2}, \frac{\pi}{2}]$, we find that

$$
K(\sinh \alpha, \sinh \gamma, T) = \frac{1}{\pi \sqrt{\cosh \alpha\, \cosh \gamma}} \left\{ \frac{\sin[(\alpha - \gamma)T]}{(\alpha - \gamma)} - \frac{\sin[(2\beta - \alpha - \gamma)T]}{(2\beta - \alpha - \gamma)} \right.
$$
$$
\left. + O(1)\frac{1 - \exp[(\gamma - \alpha)T]}{(\gamma - \alpha)T} + O(1)\frac{1 - \exp[(2\beta - \alpha - \gamma)T]}{(2\beta - \alpha - \gamma)T} \right\}, \beta < \gamma \leq \alpha. \tag{2.1.42}
$$

Using the above estimate (2.1.42) in J_1, we obtain

$$
\begin{aligned}
J_1(\alpha, T) &= \frac{1}{\pi}\int_\beta^\alpha f(\sinh \gamma)\sqrt{\frac{\cosh \gamma}{\cosh \alpha}}\, \frac{\sin[(\alpha - \gamma)T]}{(\alpha - \gamma)}\, d\gamma \\
&\quad - \frac{1}{\pi}\int_\beta^\alpha f(\sinh \gamma)\sqrt{\frac{\cosh \gamma}{\cosh \alpha}}\, \frac{\sin[(2\beta - \alpha - \gamma)T]}{(2\beta - \alpha - \gamma)}\, d\gamma \\
&\quad + O(1)\int_\beta^\alpha |f(\sinh \gamma)|\sqrt{\frac{\cosh \gamma}{\cosh \alpha}}\, \frac{1 - \exp[(\gamma - \alpha)T]}{(\alpha - \gamma)T}\, d\gamma \\
&\quad + O(1)\int_\beta^\alpha |f(\sinh \gamma)|\sqrt{\frac{\cosh \gamma}{\cosh \alpha}}\, \frac{1 - \exp[(2\beta - \alpha - \gamma)T]}{(\alpha + \gamma - 2\beta)T}\, d\gamma.
\end{aligned}
\tag{2.1.43}
$$

33

Since $f(\sinh\gamma)\sqrt{\cosh\gamma} \in L(\beta,\infty)$, it follows that as $T \to \infty$, the first integral in the right side of (2.1.43) tends to $\frac{1}{2}f(\sinh\alpha - 0)$ and the second integral tends to zero. It is easily shown that the third and fourth integrals also tend to zero as $T \to \infty$. Hence,

$$\lim_{T\to\infty} J_1(\alpha, T) = \frac{1}{2}f(\sinh\alpha - 0). \tag{2.1.44}$$

Similarly, it can be proved that

$$\lim_{T\to\infty} J_2(\alpha, T) = \frac{1}{2}f(\sinh\alpha + 0). \tag{2.1.45}$$

Therefore, we conclude that

$$\lim_{T\to\infty} J(x, T) = \frac{1}{2}[f(x+0) + f(x-0)]. \tag{2.1.46}$$

Thus, at the points of continuity of $f(x)$, we obtain the expansion formula (2.1.35).

The following integral expansion formula is a generalization of formula (2.1.35) and holds under the conditions stated in Theorem 2.1.3 (cf. Moshinskii, 1989):

$$f(x) = i\int_0^\infty \frac{\tau\,\tanh\pi\tau\,Y_{i\tau}(x,\xi)}{Q'_{-\frac{1}{2}+i\tau}(i\xi)\,Q'_{-\frac{1}{2}-i\tau}(i\xi)} \left\{\int_\xi^\infty Y_{i\tau}(y,\xi)\,f(y)\,dy\right\} d\tau, \tag{2.1.47}$$

where

$$Y_\nu(x, y) = P_{\nu-\frac{1}{2}}(ix)\,Q'_{\nu-\frac{1}{2}}(iy) - P'_{\nu-\frac{1}{2}}(iy)\,Q_{\nu-\frac{1}{2}}(ix), \quad -\infty < \xi < \infty,$$

and the dash denotes differentiation with respect to the argument.

The formula (2.1.47) can be established rigorously in a manner similar to the proof of Theorem 2.1.3. The following estimates are needed for the establishment of the expansion formula (2.1.47):

$$|Y_{i\tau}(x, y)| \leq \frac{\pi}{\cosh\pi\tau}\max_\tau\left\{\left|P'_{-\frac{1}{2}+i\tau}(-iy)\right|, \left|P'_{-\frac{1}{2}+i\tau}(iy)\right|\right\} G(x), \tag{2.1.48}$$

where

$$G(x) \equiv \frac{1}{2}[P_{-\frac{1}{2}}(ix) + P_{-\frac{1}{2}}(-ix)] = \begin{cases} O(1), & x \in (0, a), \ (a > 0), \\ \\ O(1)\,x^{-\frac{1}{2}}\ln x, & x \in (a, \infty), \end{cases}$$

and as $|\nu| \to \infty$ with $|\arg \nu| \le \frac{\pi}{2}$

$$\frac{P'_{\nu-\frac{1}{2}}(i\sinh\alpha)}{P_{\nu-\frac{1}{2}}(i\sinh\alpha)} = \frac{Q'_{\nu-\frac{1}{2}}(i\sinh\alpha)}{Q_{\nu-\frac{1}{2}}(i\sinh\alpha)} + O(|\nu|^{-1}), \quad (-\infty < \alpha < \infty). \qquad (2.1.49)$$

Under the conditions of Theorem 2.1.3, another integral expansion formula holds for $\xi = 0$ at the points of continuity of $f(x)$ and is given by

$$f(x) = 2\int_0^\infty \frac{\tau \tanh \pi\tau \ R_{i\tau}(x)}{\cosh \pi\tau \left\{\dfrac{2h}{\pi} + [P'_{-\frac{1}{2}+i\tau}(0)]^2 + [h\ P_{-\frac{1}{2}+i\tau}(0)]^2\right\}} \times$$

$$\times \left\{\int_0^\infty f(y)\ R_{i\tau}(y)\ dy\right\}\ d\tau, \qquad (2.1.50)$$

where

$$R_\nu(x) = \frac{h\ P_{\nu-\frac{1}{2}}(0)}{2i}[P_{\nu-\frac{1}{2}}(ix) - P_{\nu-\frac{1}{2}}(-ix)] - \frac{P'_{\nu-\frac{1}{2}}(0)}{2}[P_{\nu-\frac{1}{2}}(ix) + P_{\nu-\frac{1}{2}}(-ix)],$$

and h is a positive real constant.

2.2 Integral expansions involving associated Legendre functions with integral superscripts

In this section, some integral expansions involving associated Legendre functions with integral superscripts are presented.

2.2.1 The generalized Mehler–Fock integral transform or Mehler–Fock integral transform of order m (m is an integer) of a function $f(x)$ is given by (cf. Sneddon, 1972, p. 416)

$$F(\tau) = \int_1^\infty f(x)\ P^{-m}_{-\frac{1}{2}+i\tau}(x)\ dx, \qquad (2.2.1)$$

while the corresponding inverse transformation formula is

$$f(x) = (-1)^m \int_0^\infty \tau \tanh \pi\tau\ P^m_{-\frac{1}{2}+i\tau}(x)\ F(\tau)\ d\tau. \qquad (2.2.2)$$

These produce the generalized Mehler–Fock integral expansion for a function. The class of functions for which this expansion theorem is valid is not yet clearly defined, although it has been used formally in solving some boundary value problems (cf. Sneddon, 1972, p. 416 and references cited therein).

2.2.2 Lebedev and Skal'skaya (1968) developed integral expansions of the forms

$$f(x) = \frac{4}{\pi} \sum_{n=0}^{[\frac{1}{2}(m-1)]} (m - \frac{1}{2} - 2n) \, \Gamma(2m - 2n) \, \Gamma(2n+1) \, \phi^m_{m-\frac{1}{2}-2n}(x) \times$$

$$\times \int_0^\infty f(y) \, \phi^m_{m-\frac{1}{2}-2n}(y) \, dy + \frac{2}{\pi} \int_0^\infty \tau \tanh \pi\tau \, \Gamma(\frac{1}{2} + m + i\tau) \, \Gamma(\frac{1}{2} + m - i\tau) \times$$

$$\times \phi^m_{i\tau}(x) \left\{ \int_0^\infty f(y) \, \phi^m_{i\tau}(y) \, dy \right\} d\tau, \quad (0 < x < \infty, m = 0, 1, 2, \ldots), \tag{2.2.3}$$

and

$$f(x) = \frac{4}{\pi} \sum_{n=0}^{[\frac{1}{2}m]-1} (m - \frac{3}{2} - 2n) \, \Gamma(2m - 2n - 1) \, \Gamma(2n+2) \, \psi^m_{m-\frac{3}{2}-2n}(x) \times$$

$$\times \int_0^\infty f(y) \, \psi^m_{m-\frac{3}{2}-2n}(y) \, dy + \frac{2}{\pi} \int_0^\infty \tau \tanh \pi\tau \, \Gamma(\frac{1}{2} + m + i\tau) \, \Gamma(\frac{1}{2} + m - i\tau) \times$$

$$\times \psi^m_{i\tau}(x) \left\{ \int_0^\infty f(y) \, \psi^m_{i\tau}(y) \, dy \right\} d\tau, \quad (0 < x < \infty, m = 0, 1, 2, \ldots), \tag{2.2.4}$$

where

$$\phi^m_\nu(x) = \frac{1}{2} \left[e^{\mp \frac{1}{2}i\pi m} \, P^{-m}_{\nu-\frac{1}{2}}(ix) + e^{\pm \frac{1}{2}i\pi m} \, P^{-m}_{\nu-\frac{1}{2}}(-ix) \right], \tag{2.2.5}$$

$$\psi^m_\nu(x) = \frac{1}{2i} \left[e^{\mp \frac{1}{2}i\pi m} \, P^{-m}_{\nu-\frac{1}{2}}(ix) - e^{\pm \frac{1}{2}i\pi m} \, P^{-m}_{\nu-\frac{1}{2}}(-ix) \right], \tag{2.2.6}$$

where the upper and lower signs occur when $x > 0$ and $x < 0$ respectively, and $[x]$ denotes the integral part of x. Empty sums appearing in the formula (2.2.3) and (2.2.4) for $m = 0$ and $m = 1$ respectively are assumed to be equal to zero, and in these cases the expansions contain only the integral terms.

Expansions (2.2.3) and (2.2.4) are useful in solving some boudary value problems of mathematical physics and the theory of elasticity involving hyperboloids of revolution of one sheet. These formulae are generalizations of the earlier formulae (2.1.3) and (2.1.4) discussed in Section 2.1. The integral expansions (2.2.3) and (2.2.4) are proved in the following theorem.

Theorem 2.2.1

Let $f(x)$ be a given function defined on $(0, \infty)$, satisfying the following conditions:

(i) $f(x)$ is piecewise continuous and has bounded variation in $(0, \infty)$,

(ii) $f(x) \in L(0, a)$, $f(x) \, x^{-\frac{1}{2}} \ln(1 + x) \in L(a, \infty)$, $(a > 0)$.

Then $f(x)$ can be represented by the formula (2.2.3) or (2.2.4) at the points of continuity.

Further, if the function $f(x)$ is defined on the interval $(-\infty, \infty)$ and satisfies the conditions:

(i) $f(x)$ is piecewise continuous and has bounded variation in $(-\infty, \infty)$,

(ii) $f(x) |x|^{-\frac{1}{2}} \ln(1 + |x|) \in L(-\infty, -a)$, $f(x) \, x^{-\frac{1}{2}} \ln(1 + x) \in L(a, \infty)$, $a > 0$,

then $f(x)$ can be represented by an analogous formula involving $\phi_\nu^m(x)$ and $\psi_\nu^m(x)$. This formula can easily be obtained by using the formulae (2.2.3) and (2.2.4) in which $f(x)$ is replaced by the following combination of even and odd functions

$$ f(x) = \frac{1}{2}[f(x) + f(-x)] + \frac{1}{2}[f(x) - f(-x)]. $$

Proof:

Let us consider the integral

$$ J(x, T) = \frac{2}{\pi} \int_0^T \tau \tanh \pi\tau \; \Gamma(\frac{1}{2} + m + i\tau) \; \Gamma(\frac{1}{2} + m - i\tau) \; \phi_{i\tau}^m(x) \times $$

$$ \times \left\{ \int_0^\infty f(y) \, \phi_{i\tau}^m(y) \, dy \right\} d\tau, \; 0 < x < \infty, \; T > 0. \tag{2.2.7} $$

From the estimate (1.2.34), it follows that the inner integral in the right side of (2.2.7) has the estimate

$$ O(1) \int_0^a |f(y)| \, dy + O(1) \int_a^\infty |f(y)| y^{-\frac{1}{2}} \, \ln(1 + y) \, dy. $$

Therefore, by condition (ii) of the theorem, it converges absolutely and uniformly on τ in any interval $(0, T)$. Further, the uniform convergence implies that we can change the order of integration and write $J(x, T)$ in the form

$$ J(x, T) = \int_0^\infty K(x, y, T) \, f(y) \, dy, \tag{2.2.8} $$

37

where

$$K(x,y,T) = \frac{2}{\pi} \int_0^T \tau \, \tanh \pi\tau \, \Gamma(\frac{1}{2}+m+i\tau) \, \Gamma(\frac{1}{2}+m-i\tau) \, \phi_{i\tau}^m(x) \, \phi_{i\tau}^m(y) \, d\tau. \quad (2.2.9)$$

Since the integrand in the relation (2.2.9) is an even function of τ, (2.2.9) can be written as

$$K(x,y,T) = -\frac{1}{\pi i} \int_{-iT}^{iT} \nu \, \tan \pi\nu \, \Gamma(\frac{1}{2}+m+\nu) \, \Gamma(\frac{1}{2}+m-\nu) \, \phi_\nu^m(x) \, \phi_\nu^m(y) \, d\nu. \quad (2.2.10)$$

Now, the formula (cf. Erdélyi et al. 1953, p. 140(3))

$$\pi \, \tan \pi\nu \, P_{\nu-\frac{1}{2}}^{-m}(z) = Q_{-\nu-\frac{1}{2}}^{-m}(z) - Q_{\nu-\frac{1}{2}}^{-m}(z) \qquad (2.2.11)$$

yields

$$\pi \, \tan \pi\nu \, \phi_\nu^m(x) = \omega_{-\nu}^m(x) - \omega_\nu^m(x), \qquad (2.2.12)$$

where $\omega_\nu^m(x)$ is defined by the expression (1.2.29).

Using the relation (2.2.12), the representation (2.2.10) reduces to any one of the following two forms

$$K(x,y,T) = \frac{2}{\pi^2 i} \int_{-iT}^{iT} \nu \, \Gamma(\frac{1}{2}+m+\nu) \, \Gamma(\frac{1}{2}+m-\nu) \, \omega_\nu^m(x) \, \phi_\nu^m(y) \, d\nu, \; y \le x,$$

$$K(x,y,T) = \frac{2}{\pi^2 i} \int_{-iT}^{iT} \nu \, \Gamma(\frac{1}{2}+m+\nu) \, \Gamma(\frac{1}{2}+m-\nu) \, \phi_\nu^m(x) \, \omega_\nu^m(y) \, d\nu, \; y \ge x,$$

$$(2.2.13)$$

where x is fixed and greater than zero.

The integrands in the relations (2.2.13) are analytic functions of the complex varible ν, and have no singularity in the semi-plane Re $\nu \ge 0$, except for the finite number of poles

$$\nu = m - \frac{1}{2} - 2n, \quad n = 0, 1, 2, \dots, [\frac{1}{2}(m-1)]. \qquad (2.2.14)$$

Completing the contour of integration in (2.2.13) with an arc Γ_T of radius $T > m$ situated in the semi-plane Re $\nu \ge 0$ and applying the residue theorem, we obtain

$$K(x,y,T) = \frac{2}{\pi^2 i} \int_{\Gamma_T} \nu \, \Gamma(\tfrac{1}{2} + m + \nu) \, \Gamma(\tfrac{1}{2} + m - \nu) \, \omega_\nu^m(x) \, \phi_\nu^m(y) \, d\nu$$

$$-\frac{4}{\pi} \sum_{n=0}^{[\frac{1}{2}(m-1)]} (m - \frac{1}{2} - 2n) \, \Gamma(2m - 2n) \, \Gamma(2n+1) \, \phi_{m-\frac{1}{2}-2n}^m(x) \, \phi_{m-\frac{1}{2}-2n}^m(y), \quad (y \leq x),$$

and

$$K(x,y,T) = \frac{2}{\pi^2 i} \int_{\Gamma_T} \nu \, \Gamma(\tfrac{1}{2} + m + \nu) \, \Gamma(\tfrac{1}{2} + m - \nu) \, \phi_\nu^m(x) \, \omega_\nu^m(y) \, d\nu$$

$$-\frac{4}{\pi} \sum_{n=0}^{[\frac{1}{2}(m-1)]} (m - \frac{1}{2} - 2n) \, \Gamma(2m - 2n) \, \Gamma(2n+1) \, \phi_{m-\frac{1}{2}-2n}^m(x) \, \phi_{m-\frac{1}{2}-2n}^m(y), \quad (y \geq x).$$

$$(2.2.15)$$

Then equation (2.2.8) can be written as

$$J(x,T) = \int_0^x K(x,y,T) \, f(y) \, dy + \int_x^\infty K(x,y,T) \, f(y) \, dy$$

$$-\frac{4}{\pi} \sum_{n=0}^{[\frac{1}{2}(m-1)]} (m - \frac{1}{2} - 2n) \Gamma(2m - 2n) \Gamma(2n+1) \phi_{m-\frac{1}{2}-2n}^m(x) \int_0^\infty \phi_{m-\frac{1}{2}-2n}^m(y) f(y) dy.$$

$$(2.2.16)$$

Let us now investigate the behaviour of $J(x,T)$ as $T \to \infty$. Using the estimates (1.2.37), (1.2.38) and the asymptotic formulae for the gamma functions, as $|\nu| \to \infty$ with $|\arg \nu| \leq \frac{\pi}{2}$, it follows that

$$\pi^{-1} \nu \, \Gamma(\tfrac{1}{2} + m + \nu) \, \Gamma(\tfrac{1}{2} + m - \nu) \, \omega_\nu^m(\sinh \alpha) \, \phi_\nu^m(\sinh \alpha') =$$

$$= \frac{\left\{ e^{-(\alpha-\alpha')\nu} + e^{-(\alpha+\alpha')\nu} + e^{-(\alpha-\alpha')\nu} O(|\nu|^{-1}) + e^{-(\alpha+\alpha')\nu} O(|\nu|^{-1}) \right\}}{4\sqrt{\cosh \alpha \, \cosh \alpha'}}, \quad \alpha' \leq \alpha,$$

and

$$\pi^{-1} \nu \, \Gamma(\tfrac{1}{2} + m + \nu) \, \Gamma(\tfrac{1}{2} + m - \nu) \, \phi_\nu^m(\sinh \alpha) \, \omega_\nu^m(\sinh \alpha') =$$

$$= \frac{\left\{ e^{-(\alpha'-\alpha)\nu} + e^{-(\alpha'+\alpha)\nu} + e^{-(\alpha'-\alpha)\nu} O(|\nu|^{-1}) + e^{-(\alpha'+\alpha)\nu} O(|\nu|^{-1}) \right\}}{4\sqrt{\cosh \alpha \, \cosh \alpha'}}, \quad \alpha \leq \alpha'. \quad (2.2.17)$$

Assuming that $\nu = T e^{i\phi}$ ($-\frac{\pi}{2} \leq \phi \leq \frac{\pi}{2}$) holds along the arc Γ_T and using the estimates (2.2.17), it is found that

$$K(\sinh\alpha, \sinh\alpha', T) = \frac{1}{\sqrt{\cosh\alpha\ \cosh\alpha'}}\left\{\frac{\sin(\alpha-\alpha')T}{\pi(\alpha-\alpha')} + \frac{\sin(\alpha+\alpha')T}{\pi(\alpha+\alpha')}\right.$$

$$\left. +O(1)\frac{1-e^{-(\alpha-\alpha')T}}{(\alpha-\alpha')T} + O(1)\frac{1-e^{-(\alpha+\alpha')T}}{(\alpha+\alpha')T}\right\}, \quad \alpha' \le \alpha,$$

and

$$K(\sinh\alpha, \sinh\alpha', T) = \frac{1}{\sqrt{\cosh\alpha\ \cosh\alpha'}}\left\{\frac{\sin(\alpha'-\alpha)T}{\pi(\alpha'-\alpha)} + \frac{\sin(\alpha'+\alpha)T}{\pi(\alpha'+\alpha)}\right.$$

$$\left. +O(1)\frac{1-e^{-(\alpha'-\alpha)T}}{(\alpha'-\alpha)T} + O(1)\frac{1-e^{-(\alpha'+\alpha)T}}{(\alpha'+\alpha)T}\right\}, \quad \alpha \le \alpha'. \tag{2.2.18}$$

Now putting $x = \sinh\alpha$, $y = \sinh\alpha'$ in the relation (2.2.16) and then using the estimates (2.2.18), the first integral in the right side of (2.2.16) reduces to the form

$$\int_0^x K(x,y,T)\ f(y)\ dy = \int_0^\alpha K(\sinh\alpha, \sinh\alpha', T)\ f(\sinh\alpha')\ \cosh\alpha'\ d\alpha'$$

$$= \frac{1}{\pi}\int_0^\alpha f(\sinh\alpha')\left(\frac{\cosh\alpha'}{\cosh\alpha}\right)^{\frac{1}{2}}\frac{\sin(\alpha-\alpha')T}{(\alpha-\alpha')}\ d\alpha'$$

$$+\frac{1}{\pi}\int_0^\alpha f(\sinh\alpha')\left(\frac{\cosh\alpha'}{\cosh\alpha}\right)^{\frac{1}{2}}\frac{\sin(\alpha+\alpha')T}{(\alpha+\alpha')}\ d\alpha'$$

$$+O(1)\int_0^\alpha |f(\sinh\alpha')|\left(\frac{\cosh\alpha'}{\cosh\alpha}\right)^{\frac{1}{2}}\frac{1-e^{-(\alpha-\alpha')T}}{(\alpha-\alpha')T}\ d\alpha'$$

$$+O(1)\int_0^\alpha |f(\sinh\alpha')|\left(\frac{\cosh\alpha'}{\cosh\alpha}\right)^{\frac{1}{2}}\frac{1-e^{-(\alpha+\alpha')T}}{(\alpha+\alpha')T}\ d\alpha'. \tag{2.2.19}$$

From the conditions imposed on the function $f(x)$, we find that

$$\lim_{T\to\infty}\frac{1}{\pi}\int_0^\alpha f(\sinh\alpha')\left(\frac{\cosh\alpha'}{\cosh\alpha}\right)^{\frac{1}{2}}\frac{\sin(\alpha-\alpha')T}{(\alpha-\alpha')}\ d\alpha' = \frac{1}{2}f(\sinh\alpha - 0). \tag{2.2.20}$$

The other integrals in the right side of (2.2.19) tend to zero as $T \to \infty$. Hence,

$$\lim_{T\to\infty}\int_0^x K(x,y,T)\ f(y)\ dy = \frac{1}{2}f(x-0). \tag{2.2.21}$$

Similarly, we can show that

$$\lim_{T\to\infty}\int_x^\infty K(x,y,T)\ f(y)\ dy = \frac{1}{2}f(x+0). \tag{2.2.22}$$

40

Therefore, using the relations (2.2.21) and (2.2.22), we get

$$\lim_{T \to \infty} J(x,T) = \frac{1}{2}[f(x+0) + f(x-0)]$$

$$-\frac{4}{\pi} \sum_{n=0}^{[\frac{1}{2}(m-1)]} (m - \frac{1}{2} - 2n)\Gamma(2m - 2n)\Gamma(2n+1)\phi_{m-\frac{1}{2}-2n}^{m}(x) \int_{0}^{\infty} \phi_{m-\frac{1}{2}-2n}^{m}(y)f(y)dy,$$

$$(2.2.23)$$

which proves the validity of the expansion formula (2.2.3). In a similar fashion, the formula (2.2.4) can also be established.

Example 2.2.1

As an example, the expansion formula (2.2.3) can be used to obtain the following expansion formula for the function $(x^2 + 1)^{-\frac{\lambda}{2}}$ $(\lambda > \frac{1}{2})$.

$$(x^2 + 1)^{-\frac{\lambda}{2}} = \frac{2^{m-1}}{\pi \ \Gamma(\frac{1}{2}\lambda + \frac{1}{2}m) \ \Gamma(\frac{1}{2}\lambda - \frac{1}{2}m)} \times$$

$$\times \left\{ 2 \sum_{n=0}^{[\frac{1}{2}(m-1)]} (m - \frac{1}{2} - 2n) \ \Gamma(n + \frac{1}{2}) \ \Gamma(m - n) \ \Gamma(\frac{\lambda}{2} - \frac{1}{2} + \frac{m}{2} - n) \times \right.$$

$$\times \Gamma(\frac{\lambda}{2} - \frac{m}{2} + n) \ \phi_{m-\frac{1}{2}-2n}^{m}(x) + \int_{0}^{\infty} \tau \tanh \pi \tau \ \Gamma(\frac{1}{4} + \frac{m}{2} + \frac{i\tau}{2}) \times$$

$$\left. \times \Gamma(\frac{1}{4} + \frac{m}{2} - \frac{i\tau}{2}) \ \Gamma(\frac{\lambda}{2} - \frac{1}{4} + \frac{i\tau}{2}) \ \Gamma(\frac{\lambda}{2} - \frac{1}{4} - \frac{i\tau}{2}) \ \phi_{i\tau}^{m}(x) \ d\tau \right\},$$

$$(0 \le x < \infty, \ \lambda > \frac{1}{2}, \ m = 0, 1, 2, \ldots). \qquad (2.2.24)$$

2.2.3 Belova and Ufliand (1967) developed formally an integral expansion formula from consideration of the following boundary value problem. Consider the solution of the partial differential equation

$$\frac{\partial}{\partial x}\left((x^2 - 1) \ \frac{\partial u}{\partial x}\right) - \frac{\mu^2}{x^2 - 1} u = \frac{\partial u}{\partial t}, \ (\mu \text{ is a real number}) \qquad (2.2.25)$$

with the boundary conditions

$$u(a,t) = 0, \ u(\infty, t) < \infty \text{ and } u(x,0) = f(x), \ (a > 1). \qquad (2.2.26)$$

41

Using the Laplace transform, the solution of the above boundary value problem is obtained as

$$u(x,t) = e^{-i\pi\mu} \int_a^\infty f(y) \left\{ \frac{1}{2\pi i} \int_{c-i\infty}^{c+i\infty} \frac{\Gamma(\nu - \mu + 1)}{\Gamma(\nu + \mu + 1)} \frac{G(x,y,p)}{Q_\nu^\mu(a)} e^{pt} \, dp \right\} dy, \quad (2.2.27)$$

with

$$G(x,y,p) = \begin{cases} u_1(x) \, u_2(y), & x \le y, \\ u_1(y) \, u_2(x), & x \ge y, \end{cases} \quad (2.2.28)$$

where

$$u_1(x) = Q_\nu^\mu(a) \, P_\nu^\mu(x) - P_\nu^\mu(a) \, Q_\nu^\mu(x), \quad u_2(x) = Q_\nu^\mu(x),$$

and

$$\nu = -\frac{1}{2} + \sqrt{p + \frac{1}{4}} \quad \left(\text{Re} \sqrt{p + \frac{1}{4}} > 0 \right). \quad (2.2.29)$$

Applying the third boundary condition of (2.2.26), evaluating the complex integral with a cut along the segment $(-\infty, -\frac{1}{4})$ of the real axis in the complex p-plane and changing the order of integration, equation (2.2.27) produces the integral expansion formula

$$f(x) = \pi \int_0^\infty \frac{\tau \sinh \pi\tau \, w_\tau^\mu(x)}{\left| \Gamma(\frac{1}{2} + i\tau + \mu) \right|^2 \left| Q_{-\frac{1}{2}+i\tau}^\mu(a) \right|^2} \left\{ \int_a^\infty f(y) \, w_\tau^\mu(y) \, dy \right\} d\tau, \quad (2.2.30)$$

where

$$w_\tau^\mu(x) = \frac{e^{-i\pi\mu}}{\cos \pi(\mu - i\tau)} \left[Q_{-\frac{1}{2}+i\tau}^\mu(a) \, P_{-\frac{1}{2}+i\tau}^\mu(x) - P_{-\frac{1}{2}+i\tau}^\mu(a) \, Q_{-\frac{1}{2}+i\tau}^\mu(x) \right].$$

For $a = 1$, the expansion formula (2.2.30) reduces to the form

$$f(x) = \frac{1}{2} \int_0^\infty \tau \sinh \pi\tau \left| \Gamma(\frac{1}{2} + i\tau - \mu) \right|^2 P_{-\frac{1}{2}+i\tau}^\mu(x) \left\{ \int_1^\infty f(y) \, P_{-\frac{1}{2}+i\tau}^\mu(y) \, dy \right\} d\tau.$$
$$(2.2.31)$$

Another integral expansion formula can be obtained from formula (2.2.30) for $\mu = m$ $(m = 0, 1, 2, \ldots)$ and is presented by

$$f(x) = (-1)^m \int_0^\infty \frac{\Gamma(\frac{1}{2} + i\tau - m)}{\Gamma(\frac{1}{2} + i\tau + m)} \frac{\tau \tanh \pi\tau \, w_\tau^m(x)}{|Q_{-\frac{1}{2}+i\tau}^m(a)|^2} \left\{ \int_a^\infty f(y) \, w_\tau^m(y) \, dy \right\} d\tau.$$
$$(2.2.32)$$

42

For $a = 1$ and $\mu = m$, formulae (2.2.31) and (2.2.32) produce the generalized Mehler–Fock integral expansion formula (2.2.2).

These expansion formulae can be used to obtain solutions of some boundary value problems of electrostatics, heat conductivity, etc., and also some problems in the theory of elasticity, e.g. torsion of a complete toroidal segment under the action of forces applied to the spherical parts of its surface (cf. Belova and Ufliand, 1967)

2.2.4 Belova and Ufliand (1970) considered the problem of torsion of a two-sheeted hyperboloid of revolution truncated at its top by an ellipsoidal surface. To solve this torsional problem, the following integral expansion theorem has been established.

Theorem 2.2.2

Let $f(x)$ be defined on (a, ∞) $(a > 1)$ satisfying the conditions :

(i) $f(x)$ is piecewise continuous and has bounded variation in (a, ∞),

(ii) $|f(x)| \, x^{-\frac{1}{2}} \ln x \in L(a, \infty)$.

Then $f(x)$ has an integral expansion given by

$$\frac{1}{2}[f(x+0) + f(x-0)] = (-1)^m \int_0^\infty \frac{\Gamma(\frac{1}{2} + i\tau - m)}{\Gamma(\frac{1}{2} + i\tau + m)} \times$$

$$\times \frac{\tau \tanh \pi\tau \; \omega_\nu(x)}{|AQ_\nu^m(a) + BQ_\nu^{m\prime}(a)|^2} \left\{ \int_a^\infty f(y) \, \omega_\nu(y) \, dy \right\} d\tau, \tag{2.2.33}$$

where

$$\omega_\nu(x) = \left[AQ_\nu^m(a) + BQ_\nu^{m\prime}(a) \right] P_\nu^m(x) - \left[AP_\nu^m(a) + BP_\nu^{m\prime}(a) \right] Q_\nu^m(x),$$

A and B are certain real constants, $m = 0, 1, 2, \ldots, \nu = -\frac{1}{2} + i\tau$, and

$$\left.\begin{array}{l}
P_\nu^{m\prime}(x) \equiv \dfrac{d}{dx} P_\nu^m(x) = \dfrac{1}{\sqrt{x^2 - 1}} \, P_\nu^{m+1}(x) + \dfrac{mx}{(x^2 - 1)} \, P_\nu^m(x), \\[4mm]
\\
Q_\nu^{m\prime}(x) \equiv \dfrac{d}{dx} Q_\nu^m(x) = \dfrac{1}{\sqrt{x^2 - 1}} \, Q_\nu^{m+1}(x) + \dfrac{mx}{(x^2 - 1)} \, Q_\nu^m(x).
\end{array}\right\} \tag{2.2.34}$$

43

Proof:

Let us consider the integral

$$J(x,T) = (-1)^m \int_0^T \frac{\Gamma(\frac{1}{2}+i\tau-m)}{\Gamma(\frac{1}{2}+i\tau+m)} \frac{\tau \tanh \pi\tau \, \omega_\nu(x)}{|AQ_\nu^m(a) + Q_\nu^{m'}(a)|^2} \left\{ \int_a^\infty f(y) \, \omega_\nu(y) \, dy \right\} d\tau.$$
(2.2.35)

The inner integral in the representation (2.2.35) is a continuous function of τ and it follows from the estimates (1.2.20), (1.2.21) and

$$\left| AQ_\nu^m(a) + BQ_\nu^{m'}(a) \right| \le \cosh \pi\tau \, O(1), \quad \left| AP_\nu^m(a) + BP_\nu^{m'}(a) \right| \le \cosh \pi\tau \, O(1),$$

that

$$\int_a^\infty |f(y) \, \omega_\nu(y)| \, dy \le O(1) \int_a^\infty |f(y)| y^{-\frac{1}{2}} \ln y \, dy.$$
(2.2.36)

This implies that the inner integral in (2.2.35) converges uniformly and the double integral $J(x,T)$ has a meaning. We note here that $AQ_\nu^m(a) + BQ_\nu^{m'}(a) \ne 0$ for Re $\nu \ge \frac{1}{2}$. Due to uniform convergence, we can change the order of integration in the representation (2.2.35) and write $J(x,T)$ as

$$J(x,T) = (-1)^m \int_a^\infty K(x,y,T) \, f(y) \, dy,$$
(2.2.37)

where

$$K(x,y,T) = \int_0^T \frac{\Gamma(\frac{1}{2}+i\tau-m)}{\Gamma(\frac{1}{2}+i\tau+m)} \frac{\tau \tanh \pi\tau \, \omega_\nu(x) \, \omega_\nu(y)}{|AQ_\nu^m(a) + BQ_\nu^{m'}(a)|^2} \, d\tau.$$
(2.2.38)

Since the integrand in (2.2.38) is an even function of τ, introducing a new variable $i\tau = s$ and using the relation (2.1.9), we obtain

$$\left.\begin{aligned}
K(x,y,T) &= \frac{1}{\pi i} \int_{-iT}^{iT} \frac{\Gamma(\frac{1}{2}+s-m)}{\Gamma(\frac{1}{2}+s+m)} \frac{s \, Q_{s-\frac{1}{2}}^m(x) \, \omega_{s-\frac{1}{2}}(y)}{AQ_{s-\frac{1}{2}}^m(a) + BQ_{s-\frac{1}{2}}^{m'}(a)} \, ds, \quad (x \ge y), \\[2ex]
K(x,y,T) &= \frac{1}{\pi i} \int_{-iT}^{iT} \frac{\Gamma(\frac{1}{2}+s-m)}{\Gamma(\frac{1}{2}+s+m)} \frac{s \, Q_{s-\frac{1}{2}}^m(y) \, \omega_{s-\frac{1}{2}}(x)}{AQ_{s-\frac{1}{2}}^m(a) + BQ_{s-\frac{1}{2}}^{m'}(a)} \, ds, \quad (x \le y).
\end{aligned}\right\}$$
(2.2.39)

Since the singularities of $\Gamma(\frac{1}{2}+s-m)$ at the points $s = m-\frac{1}{2}, m-\frac{3}{2}, \ldots$ are cancelled by the zeros of the function $\omega_{s-\frac{1}{2}}$, the integrands in $K(x,y,T)$ are regular in s in the

44

half-plane Re $s \geq 0$. Hence, the integration over the imaginary axis can be replaced by integration over the half-circle Γ_T where $s = Te^{i\phi}$ with $-\frac{\pi}{2} \leq \phi \leq \frac{\pi}{2}$.

Using the estimates (1.2.35), (1.2.36) and (1.1.12), it is found that for $x \geq y$, as $|s| \to \infty$ with $|\arg s| \leq \frac{\pi}{2}$,

$$K(x,y,T) = \frac{1}{2\pi i} \frac{(-1)^m}{\sqrt{\sinh\alpha \, \sinh\gamma}} \int_{\Gamma_T} \left[e^{-s(\alpha-\gamma)} - (-1)^k \, e^{-s(\alpha+\gamma-\alpha_0)} \right.$$

$$\left. + e^{-s(\alpha-\gamma)}\sqrt{\cosh\gamma} \, O(|s|^{-1}) + e^{-s(\alpha+\gamma-2\alpha_0)} O(|s|^{-1}) + e^{-s(\alpha+\gamma)} O(|s|^{-1}) \right] ds$$

$$= \frac{1}{\pi} \frac{(-1)^m}{\sqrt{\sinh\alpha \, \sinh\gamma}} \left[\frac{\sin T(\alpha-\gamma)}{(\alpha-\gamma)} - (-1)^k \frac{\sin T(\alpha+\gamma-2\alpha_0)}{(\alpha+\gamma-2\alpha_0)} \right.$$

$$+ O(1) \left\{ \sqrt{\cosh\gamma} \, \frac{1-e^{-(\alpha-\gamma)T}}{(\alpha-\gamma)T} + \frac{1-e^{-(\alpha+\gamma-2\alpha_0)T}}{(\alpha+\gamma-2\alpha_0)T} + \left. \frac{1-e^{-(\alpha+\gamma)T}}{(\alpha+\gamma)T} \right\} \right], \quad (2.2.40)$$

where $x = \cosh\alpha$, $y = \cosh\gamma$, $a = \cosh\alpha_0$, $k = 0$ for $B = 0$ and 1 for $B \neq 0$. For $x \leq y$, i.e., $\alpha \leq \gamma$ in the estimate of (2.2.40), α and γ are to be interchanged. Hence, we write $J(x,T)$ as

$$J(x,T) = (-1)^m \int_{\alpha_0}^{\alpha} K(x,y,T) \, f(y) \, dy + (-1)^m \int_{\alpha}^{\infty} K(x,y,T) \, f(y) \, dy$$

$$= J_1(x,T) + J_2(x,T), \text{ say.} \quad (2.2.41)$$

Proceeding as in the previous Section 2.1, we find that

$$\lim_{T\to\infty} J_1(x,T) = \frac{1}{2}f(x-0). \quad (2.2.42)$$

Similarly,

$$\lim_{T\to\infty} J_2(x,T) = \frac{1}{2}f(x+0). \quad (2.2.43)$$

Therefore, the expansion formula (2.2.33) is proved.

Example 2.2.2

As an example of the above expansion formula, Belova and Ufliand (1970) considered the problem of the torsion of a two-sheeted hyperboloid of revolution, truncated at its top by an ellipsoidal surface. Let (α, β, ϕ) denote prolate spheriodal coordinates connected with the cylindrical coordinates (r, z, ϕ) by

$$r = c \, \sinh\alpha \, \sin\beta, \quad z = c \, \cosh\alpha \, \cos\beta \ (c > 0).$$

45

Then we have to solve the partial differential equation

$$\left(\nabla^2 - \frac{1}{r^2}\right) u = 0, \tag{2.2.44}$$

in the region $\alpha_0 \leq \alpha < \infty$, $0 \leq \beta \leq \beta_0$ with boundary conditions on the surfaces $\alpha = \alpha_0$ and $\beta = \beta_0$. ∇^2 denotes the Laplacian whose form in the (α, β, ϕ) coordinate system is given by

$$\nabla^2 u = \frac{1}{c^2(\cosh^2 \alpha - \sin^2 \beta)} \left[\frac{1}{\cosh \alpha} \frac{\partial}{\partial \alpha} \left(\cosh \alpha \frac{\partial u}{\partial \alpha} \right) + \frac{1}{\sin \beta} \frac{\partial}{\partial \beta} \left(\sin \beta \frac{\partial u}{\partial \beta} \right) \right.$$
$$\left. + \left(\frac{1}{\sin^2 \beta} - \frac{1}{\cosh^2 \alpha} \right) \frac{\partial^2 u}{\partial \phi^2} \right].$$

It is assumed that the shear stress due to torsion, applied to the surface of the hyperboloid $\beta = \beta_0$ is given by $\tau_{\beta\phi} = h(\alpha)$. Then the boundary conditions for the displacement $u(\alpha, \beta) \equiv u_\phi(\alpha, \beta)$ are

$$u(\alpha_0, \beta) = 0, \quad \frac{\partial}{\partial \beta} \left(\frac{u}{\sin \beta} \right) (\alpha, \beta_0) = \frac{c \sqrt{\sinh^2 \alpha + \sin^2 \beta_0}}{S \sin \beta_0} h(\alpha) = f(\alpha), \text{ say}, \tag{2.2.45}$$

where S is the shear modulus.

The solution of the above boundary value problem can be represented by

$$u(\alpha, \beta) = \int_0^\infty w_\nu(\alpha) \, P_\nu^1(\cos \beta) \, F(\tau) \, d\tau, \tag{2.2.46}$$

where $w_\nu(\alpha) = Q_\nu^1(\cosh \alpha_0) \, P_\nu^1(\cosh \alpha) - P_\nu^1(\cosh \alpha_0) \, Q_\nu^1(\cosh \alpha)$, $\nu = -\frac{1}{2} + i\tau$ and $F(\tau)$ is an unknown function to be determined.

Using the boundary condition (2.2.45) in (2.2.46) and then from the expansion formula (2.2.33), the function $F(\tau)$ can be obtained as
$$F(\tau) = -\frac{\tau \, \tanh \pi\tau \, \sin^2 \beta_0}{(\frac{1}{4} + \tau^2) \, |Q_\nu^1(\cosh \alpha_0)|^2 \, [\sin^2 \beta_0 \, P_\nu^{1\prime}(\cos \beta_0) + \cos \beta_0 \, P_\nu^1(\cos \beta_0)]} \times$$
$$\times \int_{\alpha_0}^\infty f(\alpha) \, w_\nu(\alpha) \, \sinh \alpha \, d\alpha. \tag{2.2.47}$$

Hence, the solution of the above boundary value problem is determined by equation (2.2.46) with (2.2.47).

2.3 Integral expansions involving associated Legendre functions with complex superscripts

In this section some integral expansions involving associated Legendre functions with complex superscripts are presented.

2.3.1 Braaksma and Meulenbeld (1967) established integral expansion theorems involving associated Legendre functions with complex superscripts. These are to some extent generalizations of the well-known Mehler–Fock expansion theorem discussed in Sections 2.1 and 2.2. Some of them are stated here without proof. The proofs can be found in the original paper of Braaksma and Meulenbeld (1967).

Theorem 2.3.1

Let the function $f(x)$ be defined on $(1,\infty)$ and have bounded variation in the open interval $(1,\infty)$. Also let $f(x)$ satisfy the following conditions:

$$\begin{aligned}
&\text{(i)} &&f(x)(x-1)^{-\frac{1}{4}-\frac{1}{2}|\operatorname{Re}\mu|} \in L(1,a) \ (a>1), &&\operatorname{Re}\mu \neq 0,\\
&\text{(ii)} &&f(x)(x-1)^{-\frac{1}{4}}\ln(x-1) \in L(1,a), &&\operatorname{Re}\mu = 0,\\
&\text{(iii)} &&f(x)x^{-\frac{1}{2}} \in L(a,\infty),
\end{aligned}$$

where $\operatorname{Re}\mu < \frac{1}{2}$. Then

$$\frac{1}{2}[f(x+0)+f(x-0)] = \int_0^\infty \pi^{-1}\,\tau\,\sinh\pi\tau\,\Gamma(\frac{1}{2}-\mu+i\tau)\,\Gamma(\frac{1}{2}-\mu-i\tau)\times$$

$$\times P^\mu_{-\frac{1}{2}+i\tau}(x)\left\{\int_1^\infty f(y)\,P^\mu_{-\frac{1}{2}+i\tau}(y)\,dy\right\}d\tau. \qquad (2.3.1)$$

If $\mu = 0$, (2.3.1) becomes

$$\frac{1}{2}[f(x+0)+f(x-0)] = \int_0^\infty \tau\,\tanh\pi\tau\,P_{-\frac{1}{2}+i\tau}(x)\left\{\int_1^\infty f(y)\,P_{-\frac{1}{2}+i\tau}(y)\,dy\right\}d\tau. \quad (2.3.2)$$

Equation (2.3.2) is the Mehler–Fock expansion formula given in Section 2.1.

Theorem 2.3.2

Let the function $f(x)$ be defined on $(0, \infty)$ and have bounded variation in the open interval $(0, \infty)$ satisfying the conditions

$$\text{(i)} \qquad f(x) \in L(0,1),$$

$$\text{(ii)} \quad x^{|\text{Re}\,\mu|-\frac{1}{2}} f(x) \in L(1,\infty), \quad \text{Re}\,\mu < 1.$$

Then

$$\frac{1}{2}[f(x+0) + f(x-0)] = \pi^{-1}\, x\, \sinh \pi x\, \Gamma(\frac{1}{2} - \mu + ix)\, \Gamma(\frac{1}{2} - \mu - ix) \times$$

$$\times \int_1^\infty P^\mu_{-\frac{1}{2}+ix}(y) \left\{ \int_0^\infty f(\tau)\, P^\mu_{-\frac{1}{2}+i\tau}(y)\, d\tau \right\} dy. \tag{2.3.3}$$

If $\mu = 0$, formula (2.3.3) becomes

$$\frac{1}{2}[f(x+0) + f(x-0)] = x\, \tanh \pi x \int_1^\infty P_{-\frac{1}{2}+ix}(y) \left\{ \int_0^\infty f(\tau)\, P_{-\frac{1}{2}+i\tau}(y)\, d\tau \right\} dy. \tag{2.3.4}$$

Equation (2.3.4) is another form of the Mehler–Fock expansion formula.

Nikolaev (1970a) established an integral expansion theorem somewhat similar to (2.3.1). This is stated here without proof.

Theorem 2.3.3

Let the function $f(x)$ be defined on $(1, \infty)$ and satisfy the conditions that

(i) $f(x)$ is piecewise continuous and possesses bounded variation in $(1, \infty)$,

(ii) $|f(x)|\, P^{-\sigma}_{-\frac{1}{2}}(x) \in L(1,\infty), \ (-\frac{1}{2} < \sigma < \infty)$.

Then

$$\frac{1}{2}[f(x+0) + f(x-0)] = \int_0^\infty \pi^{-1}\, \tau\, \sinh \pi\tau\, \Gamma(\frac{1}{2} + \mu + i\tau)\, \Gamma(\frac{1}{2} + \mu - i\tau)\, P^{-\mu}_{-\frac{1}{2}+i\tau}(x) \times$$

$$\times \left\{ \int_1^\infty f(y)\, P^{-\mu}_{-\frac{1}{2}+i\tau}(y)\, dy \right\} d\tau, \ \text{Re}\,\mu > -\frac{1}{2}. \tag{2.3.5}$$

The proof of this theorem can be established in a manner similar to the proofs of the previous theorems. The expansion formula (2.3.5) is in fact the same as (2.3.1) if μ is replaced by $-\mu$.

2.3.2 As an illustration of the expansion formula (2.3.5), the following integral expansion of the function $e^{-a\cosh\alpha}\sinh^\mu\alpha$ $(a > 0, -\frac{1}{2} < \operatorname{Re}\mu < 1)$ has been obtained by Nikolaev (1970a).

Example 2.3.1

$$e^{-a\cosh\alpha}\sinh^\mu\alpha = \frac{\sqrt{2}}{\pi\sqrt{\pi}}\, a^{-\frac{1}{2}-\mu}\int_0^\infty \tau\,\sinh\pi\tau\,\Gamma(\frac{1}{2}+\mu+i\tau)\,\Gamma(\frac{1}{2}+\mu-i\tau)\times$$
$$\times K_{i\tau}(a)\,P_{-\frac{1}{2}+i\tau}^{-\mu}(\cosh\alpha)\,d\tau. \tag{2.3.6}$$

The proof follows from the following integral representation of $K_{i\tau}(a)$ (cf. Erdélyi et al. 1954, vol. 2, p. 323(11)):

$$K_{i\tau}(a) = \sqrt{\frac{\pi}{2}}\,a^{\mu+\frac{1}{2}}\int_0^\infty e^{-a\cosh\alpha}\,\sinh^{1+\mu}\alpha\,P_{-\frac{1}{2}+i\tau}^{-\mu}(\cosh\alpha)\,d\alpha, \quad (a > 0,\ \operatorname{Re}\mu < 1),$$

and then use of formula (2.3.5).

Also, Nikolaev (1970b) used the integral expansion formula (2.3.5) to obtain closed form solutions of two linear integral equations with symmetric kernels, given as follows.

Example 2.3.2

First we consider the integral equation

$$f(x) = g(x) + \lambda\int_1^\infty \frac{(x^2-1)^{\frac{\mu}{2}}(y^2-1)^{\frac{\mu}{2}}}{(x+y)^{1+2\mu}}\,f(y)\,dy, \tag{2.3.7}$$

for $1 \le x < \infty$, $|\lambda| < \dfrac{\Gamma(2\mu+1)}{\Gamma^2(\mu+\frac{1}{2})}$, $\operatorname{Re}\mu > -\frac{1}{2}$.

A formal solution of the integral equation (2.3.7) is obtained as follows. We assume that there exist functions

$$\begin{aligned}F(\tau) &= \int_1^\infty f(y)\,P_{-\frac{1}{2}+i\tau}^{-\mu}(y)\,dy,\\ G(\tau) &= \int_1^\infty g(y)\,P_{-\frac{1}{2}+i\tau}^{-\mu}(y)\,dy.\end{aligned}\right\} \tag{2.3.8}$$

Now we multiply both sides of equation (2.3.7) by $P_{-\frac{1}{2}+i\tau}^{-\mu}(x)$ and integrate the result with respect to x from 1 to ∞. Then using the integral representation (1.2.17), we obtain

$$F(\tau) = G(\tau) + \frac{\lambda\,\Gamma(\frac{1}{2}+\mu+i\tau)\,\Gamma(\frac{1}{2}+\mu-i\tau)}{\Gamma(1+2\mu)}\,F(\tau). \tag{2.3.9}$$

From (2.3.9), it follows that

$$F(\tau) = \frac{G(\tau)}{1 - \dfrac{\lambda\, \Gamma(\frac{1}{2}+\mu+i\tau)\, \Gamma(\frac{1}{2}+\mu-i\tau)}{\Gamma(1+2\mu)}}.$$

(2.3.10)

Hence, using the inversion formula (2.3.5), the relation (2.3.10) gives the solution of the integral equation (2.3.7) as

$$f(x) = \int_0^\infty \pi^{-1} \frac{\Gamma(\frac{1}{2}+\mu+i\tau)\, \Gamma(\frac{1}{2}+\mu-i\tau)\, \tau\, \sinh\pi\tau\; G(\tau)}{1 - \dfrac{\lambda\, \Gamma(\frac{1}{2}+\mu+i\tau)\, \Gamma(\frac{1}{2}+\mu-i\tau)}{\Gamma(1+2\mu)}}\; P^{-\mu}_{-\frac{1}{2}+i\tau}(x)\; d\tau.$$

(2.3.11)

Remark:

The conditions under which the solution (2.3.11) actually defines a solution of the integral equation (2.3.7) are the following.

Let $g(x)$ be a given continuous function defined on $(1,\infty)$ with bounded variation in the open interval $(1,\infty)$ satisfying the conditions

$$\text{(i)} \quad |g(x)|\, P^{-\sigma}_{-\frac{1}{2}}(x) \;\in\; L(1,\infty), \quad (-\tfrac{1}{2} < \sigma < \infty),$$

$$\text{(ii)} \quad \tau\,|G(\tau)| \qquad \in\; L(0,\infty).$$

Then the integral on the right side of (2.3.11) exists and defines a continuous solution of the integral equation (2.3.7).

Proof:

The formula

$$2^{1+2\mu}\, \Gamma(\tfrac{1}{2}+\mu+i\tau)\, \Gamma(\tfrac{1}{2}+\mu-i\tau) = \Gamma(1+2\mu)\int_0^\infty \frac{\cos\tau t}{\cosh^{1+2\mu} t}\, dt,$$

implies the inequality

$$\left|\Gamma(\tfrac{1}{2}+\mu+i\tau)\, \Gamma(\tfrac{1}{2}+\mu-i\tau)\right| \leq \left|\Gamma(\tfrac{1}{2}+\mu)\right|.$$

Use of this in (2.3.11) produces

$$\left| \frac{\tau \, \sinh \pi\tau \, \Gamma(\tfrac{1}{2}+\mu+i\tau) \, \Gamma(\tfrac{1}{2}+\mu-i\tau) \, G(\tau) \, P^{-\mu}_{-\frac{1}{2}+i\tau}(x)}{1 - \dfrac{\lambda \, \Gamma(\tfrac{1}{2}+\mu+i\tau) \, \Gamma(\tfrac{1}{2}+\mu-i\tau)}{\Gamma(1+2\mu)}} \right|$$

$$\leq \frac{\Gamma(\tfrac{1}{2}+\sigma) \, P^{-\sigma}_{-\frac{1}{2}}(x)}{1 - |\lambda| \dfrac{\Gamma^2(\tfrac{1}{2}+\mu)}{\Gamma(1+2\mu)}} \, \tau \, |G(\tau)|, \quad -\frac{1}{2} < \sigma < \infty, \qquad (2.3.12)$$

where $|\lambda| < \dfrac{\Gamma(1+2\mu)}{\Gamma^2(\tfrac{1}{2}+\mu)}$ and $\operatorname{Re}\mu > -\tfrac{1}{2}$.

The inequality (2.3.12) and condition (ii) imply that the function $f(x)$ defined by equation (2.3.11) exists and is continuous in the interval $(0,\infty)$.

Again, it is found that

$$\lambda \int_1^\infty \frac{(x^2-1)^{\frac{\mu}{2}}(y^2-1)^{\frac{\mu}{2}}}{(x+y)^{1+2\mu}} \, f(y) \, dy = \frac{\lambda}{\pi}\int_0^\infty \frac{\tau \, \sinh \pi\tau \, \Gamma(\tfrac{1}{2}+\mu+i\tau)}{1 - \dfrac{\lambda \, \Gamma(\tfrac{1}{2}+\mu+i\tau) \, \Gamma(\tfrac{1}{2}+\mu-i\tau)}{\Gamma(1+2\mu)}} \times$$

$$\times\Gamma(\frac{1}{2}+\mu-i\tau)\left\{ \int_1^\infty \frac{(x^2-1)^{\frac{\mu}{2}}(y^2-1)^{\frac{\mu}{2}}}{(x+y)^{1+2\mu}} \, P^{-\mu}_{-\frac{1}{2}+i\tau}(y) \, dy \right\} d\tau$$

$$= -\frac{1}{\pi}\int_0^\infty \tau \, \sinh\pi\tau \, \Gamma(\frac{1}{2}+\mu+i\tau) \, \Gamma(\frac{1}{2}+\mu-i\tau) \, P^{-\mu}_{-\frac{1}{2}+i\tau}(x) \, G(\tau) \, d\tau$$

$$+\frac{1}{\pi}\int_0^\infty \frac{\tau \, \sinh\pi\tau \, \Gamma(\tfrac{1}{2}+\mu+i\tau) \, \Gamma(\tfrac{1}{2}+\mu-i\tau)}{1 - \dfrac{\lambda \, \Gamma(\tfrac{1}{2}+\mu+i\tau) \, \Gamma(\tfrac{1}{2}+\mu-i\tau)}{\Gamma(1+2\mu)}} \, P^{-\mu}_{-\frac{1}{2}+i\tau}(x) \, d\tau$$

$$= -g(x) + f(x). \qquad (2.3.13)$$

This is the integral equation (2.3.7). We have changed the order of integration in the integrals of equation (2.3.13) due to the conditions imposed on the function $g(x)$.

Example 2.3.3

We consider another integral equation

$$f(x) = g(x) + \lambda \int_1^\infty \frac{(x^2-1)^{\frac{\mu}{2}}(y^2-1)^{\frac{\mu}{2}}}{(x^2+y^2-1)^{\mu+\frac{1}{2}}} \, f(y) \, dy, \qquad (2.3.14)$$

where $1 \leq x < \infty$, $|\lambda| < \dfrac{2\Gamma(\tfrac{1}{2}+\mu)}{\Gamma^2(\tfrac{1}{4}+\tfrac{\mu}{2})}$ and $\operatorname{Re}\mu > -\dfrac{1}{2}$. Proceeding as in Example 2.3.2, using the identity (1.2.18), from the integral equation (2.3.14), we obtain

$$F(\tau) = G(\tau) + \frac{\lambda \, \Gamma(\tfrac{1}{4}+\tfrac{1}{2}\mu+\tfrac{1}{2}i\tau) \, \Gamma(\tfrac{1}{4}+\tfrac{1}{2}\mu-\tfrac{1}{2}i\tau)}{2\Gamma(\tfrac{1}{2}+\mu)} \, F(\tau). \qquad (2.3.15)$$

Then, by using the expansion formula (2.3.5), the solution of the integral equation (2.3.14) is obtained as

$$f(x) = \frac{1}{\pi} \int_0^\infty \frac{\tau \sinh \pi\tau \; \Gamma(\frac{1}{2} + \mu + i\tau) \; \Gamma(\frac{1}{2} + \mu - i\tau) \; G(\tau) \; P_{-\frac{1}{2}+i\tau}^{-\mu}(x)}{1 - \dfrac{\lambda \; \Gamma(\frac{1}{4} + \frac{1}{2}\mu + \frac{1}{2}i\tau) \; \Gamma(\frac{1}{4} + \frac{1}{2}\mu - \frac{1}{2}i\tau)}{2\,\Gamma(\frac{1}{2} + \mu)}} \, d\tau. \qquad (2.3.16)$$

As in the case of the previous example, if the function $g(x)$ satisfies the same conditions, then the solution (2.3.16) exists.

Example 2.3.4

The integral expansion formula (2.3.1) or (2.3.5) can also be used to solve a number of dual integral equations. Pathak (1978) considered five pairs of dual integral equations with trigonometric functions as kernel and obtained closed form solutions by using the integral expansion formula (2.3.1) and some properties of the associated Legendre functions. These dual integral equations with their solutions are given below:

(i)
$$\left. \begin{aligned} \int_0^\infty f(\tau) \cos x\tau \; d\tau &= f_1(x), \quad 0 \le x \le c, \\ \int_0^\infty f(\tau) \, w(\tau) \, \operatorname{cosech} \pi\tau \, \sin x\tau \; d\tau &= f_2(x), \quad c < x < \infty, \end{aligned} \right\} \qquad (2.3.17)$$

where $w(\tau) = \left\{ \Gamma(\frac{1}{2} - \mu + i\tau) \; \Gamma(\frac{1}{2} - \mu - i\tau) \right\}^{-1}$.

Solution:

$$f(\tau) = \pi^{-1} \tau \, \sinh \pi\tau \, \{w(\tau)\}^{-1} \left[\sqrt{\frac{2}{\pi}} \left\{ \Gamma\left(\frac{1}{2} - \mu\right) \right\}^{-1} \int_0^c \sinh^{1+\mu} \alpha \times \right.$$

$$\times P_{-\frac{1}{2}+i\tau}^{\mu}(\cosh \alpha) \left\{ \int_0^\alpha \frac{f_1(x)}{(\cosh \alpha - \cosh x)^{\frac{1}{2}+\mu}} \, dx \right\} d\alpha + \sqrt{2\pi} \left\{ \Gamma\left(\frac{1}{2} + \mu\right) \right\}^{-1} \times$$

$$\times \int_c^\infty \sinh^{1-\mu} \alpha \; P_{-\frac{1}{2}+i\tau}^{\mu}(\cosh \alpha) \left\{ \int_\alpha^\infty \frac{f_2(x)}{(\cosh x - \cosh \alpha)^{\frac{1}{2}-\mu}} \, dx \right\} d\alpha \right]. \qquad (2.3.18)$$

(ii)
$$\left. \begin{aligned} \int_0^\infty \tau^{-1} f(\tau) \sin x\tau \; d\tau &= f_3(x), \quad 0 \le x \le c, \\ \int_0^\infty f(\tau) \, w(\tau) \, \operatorname{cosech} \pi\tau \, \sin x\tau \; d\tau &= f_2(x), \quad c < x < \infty. \end{aligned} \right\} \qquad (2.3.19)$$

The solution of the dual integral equations (2.3.19) is the same as that given by

52

(2.3.18) with $f_1(x) = f_3'(x)$.

(iii) $\quad \int_0^\infty \tau\, f(\tau)\, \sin x\tau\, d\tau = f_4(x), \quad 0 \le x \le c,$

$\qquad \int_0^\infty f(\tau)\, \omega(\tau)\, \operatorname{cosech} \pi\tau\, \sin x\tau\, d\tau = f_2(x), \quad c < x < \infty.$ $\qquad\qquad$ (2.3.20)

The solution of the dual integral equations (2.3.20) is the same as that given by (2.3.18) with $f_1(x) = C - \int_0^x f_4(t)\, dt$ where $C = \int_0^\infty f(\tau)\, d\tau$, which can be obtained by an obvious manipulation.

(iv) $\quad \int_0^\infty f(\tau)\, \cos x\tau\, d\tau = f_1(x), \quad 0 \le x \le c,$

$\qquad \int_0^\infty \tau^{-1} f(\tau)\, \omega(\tau)\, \operatorname{cosech} \pi\tau\, \cos x\tau\, d\tau = f_5(x), \quad c < x < \infty.$ $\qquad\qquad$ (2.3.21)

The solution of the dual integral equations (2.3.21) is the same as that given by (2.3.18) with $f_2(x) = -f_5'(x)$.

(v) $\quad \int_0^\infty f(\tau)\, \cos x\tau\, d\tau = f_1(x), \quad 0 \le x \le c,$

$\qquad \int_0^\infty \tau\, f(\tau)\, \omega(\tau)\, \operatorname{cosech} \pi\tau \cos x\tau\, d\tau = f_6(x), \quad c < x < \infty.$ $\qquad\qquad$ (2.3.22)

The solution of the dual integral equation (2.3.22) is the same as that given by (2.3.18) with $f_2(x) = C - \int_x^\infty f_6(t)\, dt$.

For the details of the method of solution, the reader is referred to the original paper of Pathak (1978).

2.3.3 Moshinskii (1990) established the following integral expansion theorem related to the Mehler–Fock transform in terms of associated spherical harmonics with imaginary superscripts.

Theorem 2.3.4

Let the function $f(x)$ be defined on $(0, \infty)$ and satisfy the following conditions:

(i) $f(x)$ is piecewise continuous and has bounded variation in $(0, \infty)$,

(ii) $f(x) \in L(0, a)$, $f(x)\, x^{-\frac{1}{2}} \ln(1 + x) \in L(a, \infty)$, $(a > 0)$.

Then, at the points of continuity of $f(x)$, the following two integral expansion formulae hold:

$$f(x) = 2 \int_0^\infty \frac{\tau \sinh \pi\tau}{(\cosh^2 \pi\tau + \sinh^2 \pi\mu)} \left| \Gamma\left(\frac{1}{4} + \frac{1}{2}i\tau - \frac{1}{2}i\mu\right) \Gamma\left(\frac{1}{4} + \frac{1}{2}i\tau + \frac{1}{2}i\mu\right) \right|^2 \times$$

$$\times \Phi_1^\mu(x,\tau) \left\{ \int_0^\infty f(y) \, \Phi_1^\mu(y,\tau) \, dy \right\} d\tau, \tag{2.3.23}$$

and

$$f(x) = 2 \int_0^\infty \frac{\tau \sinh \pi\tau}{(\cosh^2 \pi\tau + \sinh^2 \pi\mu)} \left| \Gamma\left(\frac{3}{4} + \frac{1}{2}i\tau - \frac{1}{2}i\mu\right) \Gamma\left(\frac{3}{4} + \frac{1}{2}i\tau + \frac{1}{2}i\mu\right) \right|^2$$

$$\times \Phi_2^\mu(x,\tau) \left\{ \int_0^\infty f(y) \, \Phi_2^\mu(y,\tau) \, dy \right\} d\tau, \tag{2.3.24}$$

where

$$\Phi_1^\mu(x,\tau) = \frac{\exp(-\frac{\pi\mu}{2}) \, P_{-\frac{1}{2}+i\tau}^{i\mu}(ix) - \exp(\frac{\pi\mu}{2}) \, P_{-\frac{1}{2}+i\tau}^{i\mu}(-ix)}{i \, 2^{1+i\mu} \, \Gamma(\frac{1}{4} + \frac{1}{2}i\tau + \frac{1}{2}i\mu) \Gamma(\frac{1}{4} - \frac{1}{2}i\tau + \frac{1}{2}i\mu)}, \tag{2.3.25}$$

and

$$\Phi_2^\mu(x,\tau) = \frac{\exp(-\frac{\pi\mu}{2}) \, P_{-\frac{1}{2}+i\tau}^{i\mu}(ix) + \exp(\frac{\pi\mu}{2}) \, P_{-\frac{1}{2}+i\tau}^{i\mu}(-ix)}{2^{1+i\mu} \, \Gamma(\frac{3}{4} + \frac{1}{2}i\tau + \frac{1}{4}i\mu) \Gamma(\frac{3}{4} - \frac{1}{2}i\tau + \frac{1}{2}i\mu)}, \tag{2.3.26}$$

with μ a real number.

The denominators in the representations (2.3.25) and (2.3.26) were chosen so that $\Phi_n^\mu(x,\tau)$ $(n = 1, 2)$ become real on the integration paths in the formulae (2.3.23) and (2.3.24). These two expansion formulae can be considered as generalizations of the formulae (2.1.2), (2.1.3) and (2.2.3), (2.2.4) respectively.

The proof of this expansion theorem is similar to the proofs of expansion theorems discussed in Sections 2.1 and 2.2. Using the relations (1.2.14) − (1.2.16) and the estimate (1.2.27), the theorem can be proved easily.

2.4 Integral expansions involving generalized associated Legendre functions with complex superscripts

In this section, some integral expansion theorems involving generalized associated Legendre functions with complex superscripts are presented.

2.4.1 Braaksma and Meulenbeld (1967) established four expansion theorems, related to the Mehler–Fock transforms, involving generalized associated Legendre functions.

Theorem 2.4.1

Let the function $f(x)$ be defined on $(1, \infty)$ and have bounded variation in the open interval $(1, \infty)$. Also let $f(x)$ satisfy the following conditions:

$$(i) \qquad f(x)\,(x-1)^{-\frac{1}{4}-\frac{1}{2}|\mathrm{Re}\ \mu|} \ \in \ L(1,a) \quad (\mathrm{Re}\ \mu \neq 0),$$

$$(ii) \quad f(x)\,(x-1)^{-\frac{1}{4}}\ln(x-1) \ \in \ L(1,a) \quad (\mathrm{Re}\ \mu = 0),$$

$$(iii) \qquad\qquad f(x)\,x^{-1-\gamma} \ \in \ L(a,\infty),$$

for all $a > 1$ where γ is a real number such that

$$\gamma > \frac{1}{2}\ \mathrm{Re}\ \mu + \frac{1}{2}\,|\mathrm{Re}\ \nu| - 1. \tag{2.4.1}$$

Then
$$\frac{1}{2}\,[f(x+0) + f(x-0)] = \frac{1}{2\pi i} \int_{\gamma-i\infty}^{\gamma+i\infty} (2k+1)\ P_k^{\mu,\nu}(x) \times$$

$$\times \left\{ \int_1^\infty e^{i\pi\mu}\ Q_k^{-\mu,-\nu}(y)\ f(y)\ dy \right\} dk, \tag{2.4.2}$$

and
$$\frac{1}{2}[f(x+0) + f(x-0)] = \frac{1}{2\pi i} \int_{\gamma-i\infty}^{\gamma+i\infty} (2k+1)\ e^{i\pi\mu}\ Q_k^{-\mu,-\nu}(x) \times$$

$$\times \left\{ \int_1^\infty P_k^{\mu,\nu}(y)\ f(y)\ dy \right\} dk. \tag{2.4.3}$$

Theorem 2.4.2

Let the function $f(z)$ be continuous on $\mathrm{Re}\ z \geq \min(\gamma, \mathrm{Re}\ \gamma_0)$ and analytic on $\mathrm{Re}\ z > \min(\gamma, \mathrm{Re}\ \gamma_0)$ where γ is a real number satisfy the inequality (2.4.1) and γ_0 is a complex number such that

$$\mathrm{Re}\ \gamma_0 > \frac{1}{2}\ \mathrm{Re}\ \mu + \frac{1}{2}\,|\mathrm{Re}\ \nu| - 1, \tag{2.4.4}$$

with $\mathrm{Re}\ \gamma_0 \geq -\frac{1}{2}$. Also let $f(z)$ satisfy the conditions

(i) $f(z)\, z^{\frac{1}{2}+\mu} \in L(\gamma - i\infty, \gamma + i\infty)$;

(ii)

$$f(z) = \begin{cases} o(z^{-1-|\mathrm{Re}\,\mu|}), & \mathrm{Re}\,\mu < 1 \text{ with } \mu \neq 0, \\[2ex] \dfrac{o(1)}{z\,\ln z}, & \mu = 0, \\[2ex] o(z^{1-3\mu}), & \mathrm{Re}\,\mu > 1, \\[2ex] \dfrac{o(1)}{z^2\,\ln z}, & \mathrm{Re}\,\mu = 1, \end{cases}$$

as $|z| \to \infty$ on $\mathrm{Re}\,z \geq \min(\gamma, \mathrm{Re}\,\gamma_0)$.

If $\mathrm{Re}\,\gamma_0 = \gamma \geq -\frac{1}{2}$, then $f(z)$ has to satisfy a Hölder condition in a right neighbourhood of $z = \gamma_0$. Then

$$f(\gamma_0) = \frac{1}{2\pi i} \int_1^{\infty} e^{i\pi\mu}\, Q_{\gamma_0}^{-\mu,-\nu}(y) \left\{ \int_{\gamma-i\infty}^{\gamma+i\infty} (2k+1)\, P_k^{\mu,\nu}(y)\, f(k)\, dk \right\} dy. \tag{2.4.5}$$

Theorem 2.4.3

Let the function $f(z)$ be a continuous function on $\mathrm{Re}\,z \geq \omega$, and analytic on $\mathrm{Re}\,z > \omega$ where $\omega = \min\left\{\gamma, \left|\mathrm{Re}\left(\gamma_0 + \frac{1}{2}\right)\right| - \frac{1}{2}\right\}$ with γ a real number satisfying the inequality (2.4.1) and γ_0 a complex number satisfying

$$|\mathrm{Re}\,(2\gamma_0 + 1)| > \mathrm{Re}\,\mu + |\mathrm{Re}\,\nu| - 1. \tag{2.4.6}$$

Further let

$$f(z)\, z^{\frac{1}{2}-\mu} \in L(\gamma - i\infty, \gamma + i\infty),$$

and as $|z| \to \infty$ on $\mathrm{Re}\,z \geq \omega$,

$$f(z) = \begin{cases} z^{-|\mathrm{Re}(\mu-1)|}\, o(1), & \mathrm{Re}\,\mu \neq 1, \\[2ex] o\left(\dfrac{1}{\ln z}\right), & \mathrm{Re}\,\mu = 1. \end{cases}$$

Then

$$\int_1^\infty P_{\gamma_0}^{\mu,\nu}(y) \left\{ \int_{\gamma-i\infty}^{\gamma+i\infty} (2k+1) \, Q_k^{-\mu,-\nu}(y) \, f(k) \, dk \right\} dy = 0. \qquad (2.4.7)$$

Theorem 2.4.4

Let μ be a complex number with $\mathrm{Re}\ \mu < 1$. Let S be the strip $|\mathrm{Re}\ z| < a$ in the complex z-plane, and $\overline{S}$ the strip $|\mathrm{Re}\ z| \le a$, where a is a positive real number such that $Q_{-\frac{1}{2}+z}^{-\mu,-\nu}(z)$ has no poles in $\overline{S}$. Let γ be a real number and γ_0 be a complex number in S.

Suppose that $f(z)$ is an even function, analytic on S and continuous on $\overline{S}$, satisfying

$$f(z) \, z^{\frac{1}{2}-\mu} \in L(\gamma - i\infty, \gamma + i\infty),$$
$$f(z) = o(k^{\mu-\frac{1}{2}}) \text{ as } k \to \infty \text{ in } \overline{S}.$$

If $|\mathrm{Re}\ \gamma_0| = a$, then $f(z)$ has to satisfy a Hölder condition in a $\overline{S}$-neighbourhood of γ_0. Then

$$\frac{1}{2} f(\gamma_0) = \frac{1}{2\pi i} \int_1^\infty P_{-\frac{1}{2}+\gamma_0}^{\mu,\nu}(y) \left\{ \int_{\gamma-i\infty}^{\gamma+i\infty} k \, e^{i\pi\mu} \, Q_{-\frac{1}{2}+k}^{-\mu,-\nu}(y) \, f(k) \, dk \right\} dy. \qquad (2.4.8)$$

This formula also holds for $\gamma = \mathrm{Re}\ \gamma_0 = 0$, and $f(z)$ is an even function defined only on the line $\mathrm{Re}\ z = 0$ and satisfies the conditions

$$z \, f(z) \in L(0, i),$$
$$z^{\frac{1}{2}-\mu} \, f(z) \in L(i, i\infty),$$

and $f(z)$ is of bounded variation in a neighbourhood of $z = \gamma_0$. Then in the right hand side of (2.4.8), $\frac{1}{2} f(\gamma_0)$ has to be replaced by $\frac{1}{4}\{f(\gamma_0 + 0i) + f(\gamma_0 - 0i)\}$.

Theorem 2.4.5

Let $f(x)$ be a function defined on $(1, \infty)$ satisfying the following conditions:

(i) $f(x) \, (x-1)^{-\frac{1}{4}-\frac{1}{2}|\mathrm{Re}\ \mu|} \in L(1, a) \ (a > 1), \quad \mathrm{Re}\ \mu \ne 0$,

(ii) $f(x) \, (x-1)^{-\frac{1}{4}} \ln(x-1) \in L(1, a), \quad \mathrm{Re}\ \mu = 0$,

(iii) $f(x) \, x^{-\frac{1}{2}} \in L(a, \infty)$.

Also, let $f(x)$ be of bounded variation in the open interval $(1, \infty)$.

Then

$$\frac{1}{2}[f(x+0) + f(x-0)] = \pi^{-2}\, 2^{\mu-\nu-1} \int_0^\infty \tau\, \sinh 2\pi\tau\, \Gamma\left(\frac{1-\mu+\nu}{2} + i\tau\right) \times$$

$$\times \Gamma\left(\frac{1-\mu+\nu}{2} - i\tau\right) \Gamma\left(\frac{1-\mu-\nu}{2} + i\tau\right) \Gamma\left(\frac{1-\mu-\nu}{2} - i\tau\right) P^{\mu,\nu}_{-\frac{1}{2}+i\tau}(x)\times$$

$$\times \left\{\int_1^\infty f(y)\, P^{\mu,\nu}_{-\frac{1}{2}+i\tau}(y)\, dy\right\} d\tau, \tag{2.4.9}$$

where $|\mathrm{Re}\ \nu| < 1 - \mathrm{Re}\ \mu$.

Theorem 2.4.6

Let $f(x)$ be a function defined on $(0,\infty)$ having bounded variation in the open interval $(0,\infty)$. Also, let $f(x)$ satisfy the following conditions:

$$\text{(i)} \qquad f(x) \in L(0,1),$$

$$\text{(ii)} \quad f(x)\, x^{\mathrm{Re}\ \mu - \frac{1}{2}} \in L(1,\infty).$$

Then we obtain

$$\frac{1}{2}[f(x+0) + f(x-0)] = \frac{2^{\mu-\nu-1}}{\pi^3}\, \sinh 2\pi\tau\, \Gamma\left(\frac{1-\mu+\nu}{2} + ix\right) \Gamma\left(\frac{1-\mu+\nu}{2} - ix\right) \times$$

$$\times \Gamma\left(\frac{1-\mu-\nu}{2} + ix\right) \Gamma\left(\frac{1-\mu-\nu}{2} - ix\right) \int_1^\infty P^{\mu,\nu}_{-\frac{1}{2}+ix}(t) \left\{\int_0^\infty f(y)\, P^{\mu,\nu}_{-\frac{1}{2}+iy}(t)\, dy\right\} dt, \tag{2.4.10}$$

where $\mathrm{Re}\ \mu < 1$.

For proofs of the above theorems, the reader should consult the original paper of Braaksma and Meulenbeld (1967).

2.4.2 Applications of the integral expansion formula (2.4.2)

Let p be a complex number and γ a real number with

$$\gamma > \max\left\{-\frac{3}{2} - \mathrm{Re}\ p,\ \frac{1}{2}\, \mathrm{Re}(\mu - \nu) - 1\right\}$$

and $\mathrm{Re}\ (\nu + 2p) < -\frac{3}{2}$. Then, the following integral expansions can be obtained for $x > 1$.

Example 2.4.1

$$\frac{(x+1)^{\frac{1}{2}\nu}}{(x-1)^{p+\frac{1}{2}\nu+\frac{3}{2}}} = 2^{-p+\frac{1}{2}(\mu-\nu)-\frac{3}{2}} \Gamma\left(\frac{1}{2}\mu - \frac{1}{2}\nu - p - \frac{1}{2}\right) \Gamma\left(-\frac{1}{2}\mu - \frac{1}{2}\nu - p - \frac{1}{2}\right) \times$$

$$\times \frac{1}{2\pi i} \int_{\gamma-i\infty}^{\gamma+i\infty} (2k+1) \frac{\Gamma(k - \frac{\mu-\nu}{2} + 1)\, \Gamma(p+k+\frac{3}{2})}{\Gamma(k + \frac{\mu-\nu}{2} + 1)\, \Gamma(-p+k+\frac{1}{2})} \, P_k^{\mu,\nu}(x)\, dk. \qquad (2.4.11)$$

Example 2.4.2

For $p = -\frac{3}{2}$, (2.4.11) becomes

$$\left(\frac{x+1}{x-1}\right)^{\frac{\nu}{2}} = 2^{\frac{\mu-\nu}{2}} \frac{\Gamma(1 + \frac{1}{2}\mu - \frac{1}{2}\nu)\, \Gamma(1 - \frac{1}{2}\mu - \frac{1}{2}\nu)}{2\pi i} \int_{\gamma-i\infty}^{\gamma+i\infty} \frac{(2k+1)}{k(k+1)} \times$$

$$\times \frac{\Gamma(k - \frac{\mu-\nu}{2} + 1)}{\Gamma(k + \frac{\mu-\nu}{2} + 1)} \, P_k^{\mu,\nu}(x)\, dk, \qquad (2.4.12)$$

where $\gamma > \max\left\{0, \frac{1}{2}\operatorname{Re}(\mu-\nu) - 1\right\}$, $\operatorname{Re}\nu < \frac{3}{2}$ and $x > 1$.

Example 2.4.3

For $\mu = \nu$, (2.4.12) produces

$$\left(\frac{x+1}{x-1}\right)^{\frac{1}{2}\mu} = \frac{\Gamma(1-\mu)}{2\pi i} \int_{\gamma-i\infty}^{\gamma+i\infty} \frac{(2\nu+1)}{\nu(\nu+1)} \, P_\nu^\mu(x)\, d\nu, \qquad (2.4.13)$$

where $\gamma > 0$, $\operatorname{Re}\mu < \frac{3}{2}$ and $x > 1$.

Example 2.4.4

For $\mu = \nu$, (2.4.11) produces

$$\frac{(x+1)^{\frac{1}{2}\mu}}{(x-1)^{\frac{1}{2}\mu+p+\frac{3}{2}}} = \frac{\Gamma(-p-\frac{1}{2})\, \Gamma(-p-\mu-\frac{1}{2})}{2^{1+p+\frac{3}{2}}\, \pi i} \int_{\gamma-i\infty}^{\gamma+i\infty} \frac{(2\nu+1)\, \Gamma(p+\nu+\frac{3}{2})}{\Gamma(-p+\nu+\frac{1}{2})} \, P_\nu^\mu(x)\, d\nu, \qquad (2.4.14)$$

where $\gamma > -\frac{3}{2} - \operatorname{Re} p$, $\operatorname{Re}(\mu + 2p) < -\frac{3}{2}$ and $x > 1$.

Example 2.4.5

For $p = -1$, it is found from (2.4.11) that

$$\frac{(x+1)^{\frac{\nu}{2}}}{(x-1)^{\frac{1}{2}+\frac{\nu}{2}}} = \frac{\Gamma(\frac{1}{2} + \frac{\mu}{2} - \frac{\nu}{2})\, \Gamma(\frac{1}{2} - \frac{\mu}{2} - \frac{\nu}{2})}{2^{\frac{1+\mu-\nu}{2}}\, \pi i} \int_{\gamma-i\infty}^{\gamma+i\infty} \frac{\Gamma(k - \frac{\mu-\nu}{2} + 1)}{\Gamma(k + \frac{\mu-\nu}{2} + 1)} \, P_k^{\mu,\nu}(x)\, dk, \qquad (2.4.15)$$

where $\gamma > \frac{1}{2}\operatorname{Re}(\mu-\nu) - 1$, $\operatorname{Re}\nu < \frac{1}{2}$ and $x > 1$.

Example 2.4.6

For $\mu = \nu$, from (2.4.15) we have

$$\frac{(x+1)^{\frac{\mu}{2}}}{(x-1)^{\frac{1}{2}+\frac{\mu}{2}}} = \frac{\sqrt{2\pi}\ \Gamma(\frac{1}{2}-\mu)}{2\pi i} \int_{\gamma-i\infty}^{\gamma+i\infty} P_\nu^\mu(x)\ d\nu, \quad \text{Re}\ \mu < \frac{1}{2}\ \text{and}\ x > 1. \quad (2.4.16)$$

Example 2.4.7

The formula (2.4.11) can be transformed by using the relation (1.3.2) with a hypergeometric function. Then after replacing $\frac{1}{2}(\nu-\mu)$ by a, $1-\mu$ by μ and $1-x$ by $-2x$, we obtain

$$x^{-p-a-\frac{3}{2}} = \frac{\Gamma(-p-a-\frac{1}{2})\ \Gamma(\mu-p-a-\frac{3}{2})}{2\pi i} \int_{\gamma-i\infty}^{\gamma+i\infty} \frac{(2k+1)\ \Gamma(k+a+1)}{\Gamma(\mu)\ \Gamma(k-a+1)} \times$$

$$\times \frac{\Gamma(p+k+\frac{3}{2})}{\Gamma(-p+k+\frac{1}{2})}\ F(-k+a, k+a+1; \mu; -x)\ dk, \quad (2.4.17)$$

where $\gamma > \max(-\frac{3}{2} - \text{Re}\ p, -\text{Re}\ a - 1)$, $\text{Re}\ (2p+2a-\mu) < -\frac{5}{2}$ and $x > 0$.

Example 2.4.8

For $p = -1$, (2.4.17) becomes

$$x^{-a-\frac{1}{2}} = \frac{\Gamma(\frac{1}{2}-a)\ \Gamma(\mu-a-\frac{1}{2})}{\pi i} \int_{\gamma-i\infty}^{\gamma+i\infty} \frac{\Gamma(k+a+1)}{\Gamma(\mu)\Gamma(k-a+1)} \times$$

$$\times F(-k+a, k+a+1; \mu; -x)\ dk, \quad (2.4.18)$$

where $\gamma > -\text{Re}\ a - 1, \text{Re}\ (2a-\mu) < -\frac{1}{2}$ and $x > 0$.

Example 2.4.9

For $a = 0$, relation (2.4.18) reduces to

$$x^{-\frac{1}{2}} = \frac{\sqrt{\pi}\ \Gamma(\mu-\frac{1}{2})}{\pi i} \int_{\gamma-i\infty}^{\gamma+i\infty} \frac{F(-k, k+1; \mu; -x)}{\Gamma(\mu)}\ dk, \quad \text{Re}\ \mu > \frac{1}{2}\ \text{and}\ x > 0. \quad (2.4.19)$$

Applications of the integral expansion formula (2.4.3)

Example 2.4.10

Using the representation (1.3.2) and (1.3.3), the following formula can be derived (cf. Braaksma and Meulenbeld, 1967):

$$\frac{(x+1)^{-p-\frac{1}{2}\mu}}{(1-x)^{p+\frac{1}{2}\nu}}\ P_k^{-\nu-2p,-\mu-2p}(x) = \frac{\Gamma(2p+1)}{\Gamma(k+\frac{\mu+\nu}{2}+2p+1)\ \Gamma(\frac{\mu+\nu}{2}-k+2p)} \times$$

$$\times \int_1^\infty (y-1)^{-\frac{\mu}{2}}\ (y+1)^{-\frac{\nu}{2}}\ (x+y)^{-2p-1}\ P_k^{\mu,\nu}(y)\ dy, \quad (2.4.20)$$

60

where $\mathrm{Re}\,\mu < 1$, $|\mathrm{Re}\,(2k+1)| < \mathrm{Re}\,(\mu+\nu+4p+1)$ and $-1 < x < 1$.

Applying the above formula (2.4.20) and the integral expansion formula (2.4.3), we obtain that

$$\Gamma(1-2p)\,(y-1)^{\frac{\nu}{2}}\,(y+1)^{\frac{\nu}{2}}\,(1-x)^{-p-\frac{\mu}{2}}\,(x+1)^{-p-\frac{\mu}{2}}\,(x+y)^{2p-1} =$$

$$= \frac{1}{2\pi i}\int_{\gamma-i\infty}^{\gamma+i\infty}(2k+1)\,\Gamma\left(k-\frac{\mu+\nu}{2}-2p+1\right)\Gamma\left(-k-\frac{\mu+\nu}{2}-2p\right)e^{-i\pi\mu}\times$$

$$\times Q_k^{\mu,\nu}(y)\,P_k^{\nu+2p,\mu+2p}(x)\,dk, \tag{2.4.21}$$

where $-1 < x < 1$, $y > 1$, $\mathrm{Re}\,\mu > -\frac{3}{4}$, $\gamma > -\frac{1}{2}\mathrm{Re}\,\mu + \frac{1}{2}|\mathrm{Re}\,\nu| - 1$ and $|2\gamma+1| < 1 - \mathrm{Re}\,(\mu+\nu+4p)$.

Example 2.4.11

For $\mu = \nu$, (2.4.21) becomes
$$\frac{\Gamma(1-2p)\,(y^2-1)^{\frac{\mu}{2}}\,(x+y)^{2p-1}}{(1-x^2)^{p+\frac{\mu}{2}}} = \frac{1}{2\pi i}\int_{\gamma-i\infty}^{\gamma+i\infty}(2k+1)\,\Gamma(k-\mu-2p+1)\times$$

$$\times\Gamma(-k-\mu-2p)\,e^{-i\pi\mu}\,Q_k^{\mu}(y)\,P_k^{\mu+2p}(x)\,dk. \tag{2.4.22}$$

Example 2.4.12

Putting $p=0$ and replacing μ by $-\mu$ in (2.4.21), we have

$$\frac{2^{\mu-\nu}}{(x+y)}\left(\frac{1-x}{y+1}\right)^{-\frac{\nu}{2}}\left(\frac{1+x}{y-1}\right)^{\frac{\nu}{2}} = \frac{1}{2\pi i}\int_{\gamma-i\infty}^{\gamma+i\infty}(2k+1)\,\Gamma\left(-k+\frac{\mu-\nu}{2}\right)\times$$

$$\times\Gamma\left(k+\frac{\mu-\nu}{2}+1\right)e^{i\pi\mu}\,Q_k^{-\mu,-\nu}(y)\,P_k^{\nu,\mu}(x)\,dk, \tag{2.4.23}$$

where $y > 1$, $-1 < x < 1$, $\mathrm{Re}\,\mu < \frac{3}{4}$, $\gamma > -\frac{1}{2}\,\mathrm{Re}\,\mu + \frac{1}{2}|\mathrm{Re}\,\nu| - 1$ and $|2\gamma+1| < 1 + \mathrm{Re}\,(\mu-\nu)$.

Example 2.4.13

For $\mu = \nu$, the above formula (2.4.23) becomes

$$\frac{-1}{\pi\,(x+y)}\left(\frac{1+x}{1-x}\,\frac{y+1}{y-1}\right)^{\frac{\mu}{2}} = \frac{1}{2\pi i}\int_{\gamma-i\infty}^{\gamma+i\infty}\frac{(2\nu+1)}{\sin\pi\nu}\,e^{i\pi\mu}\,Q_\nu^{-\mu}(y)\,P_\nu^{\mu}(x)\,d\nu. \tag{2.4.24}$$

Using the results (cf. Erdélyi et al. 1954, p. 323 (11) and (12))

$$\int_1^\infty e^{-ay}\,(y^2-1)^{-\frac{\mu}{2}}\,P_\nu^{\mu}(y)\,dy = \sqrt{\frac{2}{\pi}}\,a^{\mu-\frac{1}{2}}\,K_{\nu+\frac{1}{2}}(a), \tag{2.4.25}$$

and

$$\int_1^\infty \left(\frac{y+1}{y-1}\right)^{\frac{\mu}{2}} e^{-ay} \, P_\nu^\mu(y) \, dy = \frac{1}{a} \, W_{\mu,\nu+\frac{1}{2}}(2a),$$ (2.4.26)

where $\operatorname{Re} a > 0$, $\operatorname{Re} \mu < 1$, $K_\nu(x)$ denotes the modified Bessel function of the second kind and $W_{\mu,\nu}(x)$ denotes the Whittaker function, the following two examples can be obtained from the integral expansion formula (2.4.3).

Example 2.4.14

$$e^{-as} \, (x^2 - 1)^{\frac{\mu}{2}} = \sqrt{\frac{2}{\pi}} \, \frac{a^{-\frac{1}{2}-\mu}}{2\pi i} \int_{\gamma-i\infty}^{\gamma+i\infty} (2\nu + 1) \, e^{-i\pi\mu} \, Q_\nu^\mu(x) \, K_{\nu+\frac{1}{2}}(a) \, d\nu,$$ (2.4.27)

where $x > 1$, $\operatorname{Re} a > 0$, $\operatorname{Re} \mu > -\frac{3}{4}$, $\gamma > -\frac{1}{2} \operatorname{Re} \mu + \frac{1}{2}|\operatorname{Re} \mu| - 1$.

Example 2.4.15

$$e^{-as} \left(\frac{x+1}{x-1}\right)^{\frac{\mu}{2}} = \frac{1}{2\pi i a} \int_{\gamma-i\infty}^{\gamma+i\infty} (2\nu + 1) \, e^{i\pi\mu} \, Q_\nu^{-\mu}(x) \, W_{\mu,\nu+\frac{1}{2}}(2a) \, d\nu,$$ (2.4.28)

for $x > 1$, $\operatorname{Re} a > 0$, $\operatorname{Re} \mu < \frac{3}{4}$, $\gamma > \frac{1}{2} \operatorname{Re} \mu + \frac{1}{2}|\operatorname{Re} \mu| - 1$.

Example 2.4.16

Mandal (1995) used the expansion formula (2.4.9) to solve five pairs of dual integral equations with trigonometric function kernels. He obtained closed form solutions of the following dual integral equations:

(i) $\quad \int_0^\infty f(\tau) \, \cos x\tau \, d\tau = f_1(x), \quad 0 \le x \le c,$

$\quad\quad \int_0^\infty f(\tau) \, \omega(\tau) \, \mathrm{cosech}\, 2\pi\tau \, \sin x\tau \, d\tau = f_2(x), \quad c < x < \infty,$ (2.4.29)

where

$$\omega(\tau) = \left[\Gamma\left(\frac{1-\mu+\nu}{2} + i\tau\right) \Gamma\left(\frac{1-\mu+\nu}{2} - i\tau\right) \Gamma\left(\frac{1-\mu-\nu}{2} + i\tau\right) \times$$

$$\times \Gamma\left(\frac{1-\mu-\nu}{2} - i\tau\right)\right]^{-1}.$$

Solution:

$$f(\tau) = \pi^{-2} \, 2^{\mu-\nu-1} \, \tau \, \sinh 2\pi\tau \, \{\omega(\tau)\}^{-1} \left[\int_0^c G_1(\alpha) \, P_{-\frac{1}{2}+i\tau}^{\mu,\nu}(\cosh \alpha) \, \sinh \alpha \, d\alpha\right.$$

$$\left. + \int_c^\infty H_1(\alpha) \, P_{-\frac{1}{2}+i\tau}^{\mu,\nu}(\cosh \alpha) \, \sinh \alpha \, d\alpha\right],$$ (2.4.30)

62

where

$$G_1(\alpha) = \pi^{-\frac{1}{2}} 2^{\frac{\mu-\nu-1}{2}} \left\{ \Gamma \left(\frac{1}{2} - \mu \right) \right\}^{-1} \sinh^{\mu} \alpha \int_0^{\alpha} (\cosh \alpha - \cosh x)^{-\mu-\frac{1}{2}} \times$$

$$\times F \left(\frac{\nu-\mu}{2}, -\frac{\nu+\mu}{2}; \frac{1}{2} - \mu; \frac{\cosh \alpha - \cosh x}{1 + \cosh \alpha} \right) f_1(x) \, dx, \tag{2.4.31}$$

and

$$H_1(\alpha) = (2\pi)^{\frac{3}{2}} 2^{\frac{\nu-\mu}{2}} \left\{ \Gamma \left(\frac{1}{2} + \mu \right) \right\}^{-1} \sinh^{-\mu} \alpha \int_{\alpha}^{\infty} (\cosh x - \cosh \alpha)^{\mu-\frac{1}{2}} \times$$

$$\times F \left(\frac{\mu-\nu}{2}, \frac{\mu+\nu}{2}; \frac{1}{2} + \mu; \frac{\cosh \alpha - \cosh x}{1 + \cosh \alpha} \right) f_2(x) \, dx. \tag{2.4.32}$$

(ii) $\displaystyle \int_0^{\infty} \tau^{-1} f(\tau) \sin x\tau \, d\tau = f_3(x), \quad 0 \le x \le c,$

$\displaystyle \int_0^{\infty} f(\tau) \, \omega(\tau) \, \mathrm{cosech}\, 2\pi\tau \, \sin x\tau \, d\tau = f_2(x), \quad c < x < \infty.$
$\left.\vphantom{\int}\right\} \tag{2.4.33}$

Solution:

The solution of the dual integral equations (2.4.33) is the same as that given by (2.4.30) with $f_1(x) = f_3'(x)$.

(iii) $\displaystyle \int_0^{\infty} \tau f(\tau) \sin x\tau \, d\tau = f_4(x), \quad 0 \le x \le c,$

$\displaystyle \int_0^{\infty} f(\tau) \, \omega(\tau) \, \mathrm{cosech}\, 2\pi\tau \, \sin x\tau \, d\tau = f_2(x), \quad c < x < \infty.$
$\left.\vphantom{\int}\right\} \tag{2.4.34}$

Solution:

The solution of the dual integral equations (2.4.34) is the same as that given by (2.4.30) with $f_1(x) = C - \int_0^x f_4(t) \, dt$ where $C = \int_0^{\infty} f(\tau) \, d\tau$.

(iv) $\displaystyle \int_0^{\infty} f(\tau) \cos x\tau \, d\tau = f_1(x), \quad 0 \le x \le c,$

$\displaystyle \int_0^{\infty} \tau^{-1} f(\tau) \, \omega(\tau) \, \mathrm{cosech}\, 2\pi\tau \, \cos x\tau \, d\tau = f_5(x), \quad c < x < \infty.$
$\left.\vphantom{\int}\right\} \tag{2.4.35}$

Solution:

The solution of the dual integral equations (2.4.35) is the same as that given by (2.4.30) with $f_2(x) = -f_5'(x)$.

(v) $\displaystyle \int_0^{\infty} f(\tau) \cos x\tau \, d\tau = f_1(x), \quad 0 \le x \le c,$

$\displaystyle \int_0^{\infty} \tau f(\tau) \, \omega(\tau) \, \mathrm{cosech}\, 2\pi\tau \, \cos x\tau \, d\tau = f_6(x), \quad c < x < \infty.$
$\left.\vphantom{\int}\right\} \tag{2.4.36}$

Solution:

The solution of the dual integral equations (2.4.36) is the same as that given by (2.4.30) with $f_2(x) = C - \int\limits_x^\infty f_6(t)\, dt$.

Chapter 3

Integral expansions related to Mehler–Fock type transforms involving associated Legendre functions

In this chapter, we present several integral expansion formulae involving associated Legendre functions in which the superscript of the associated Legendre functions appears as an integration variable. These expansion formulae are not widely known in the literature. The corresponding integral transforms generated by these integral expansion formulae are termed here Mehler–Fock type integral transforms. This nomenclature is used here to distinguish these from the transforms discussed in the last chapter, which are related to the usual Mehler–Fock transform. As examples, integral expansions of some simple functions are deduced from these integral expansion formulae. It may be noted that, in the absence of these expansion formulae, it is not obvious how to establish the integral expansions of the functions given in the examples.

3.1 Integral expansions in $(-1, 1)$

3.1.1 Integral expansion of a function in terms of associated Legendre functions $P_{-\frac{1}{2}+i\tau}^{-\mu}(\pm x)$ $(-1 < x < 1)$ in which the superscript appears as an integration variable

while the subscript $-\frac{1}{2} + i\tau$ (τ real) remains fixed, was first introduced by Felsen (1958). He formally developed such expansion formulae from a unique δ-function representation as mentioned in Chapter 1 (cf. equation (1.5)). He considered the Green's function $G(\theta, \theta'; \tau; \lambda)$ satisfying

$$\left[-\frac{d}{d\theta}\left(\sin\theta \frac{d}{d\theta} \right) + \left(\frac{1}{4} + \tau^2 \right)\sin\theta + \frac{\lambda}{\sin\theta} \right] G\left(\theta, \theta'; \tau; \lambda\right) = \delta\left(\theta - \theta'\right), \ \theta_1 \le \theta \le \theta_2,$$

$$(3.1.1)$$

with the boundary conditions at the end points θ_1, θ_2 as yet unspecified and τ is a specified real parameter. Here θ is the latitudinal coordinate of the spherical coordinate system used in connection with the study of some diffraction problems involving conical boundaries. A δ-function representation is obtained from the relation (1.4), as given by

$$\delta\left(\theta - \theta'\right) \sin \theta' = \frac{1}{2\pi i} \oint_\Gamma G\left(\theta, \theta'; \tau; \lambda\right) d\lambda, \qquad (3.1.2)$$

where the contour Γ encloses all the singularities of G in the anticlockwise sense. An integral expansion for a suitable function $f(\theta)$ in the domain $\theta_1 \le \theta \le \theta_2$ can be obtained formally by multiplying both sides of the relation (3.1.2) by $\frac{f(\theta')}{\sin \theta'}$ and integrating over θ' between θ_1 and θ_2. For various intervals in the θ-domain, different types of integral expansion formulae, related to Mehler–Fock type transforms, can be developed formally (see equation (1.5)).

We give below expressions for the Green's functions and corresponding δ-function representations for various intervals as well as boundary conditions.

Case I. $0 \le \theta \le \pi$

In this case, the end points $\theta = 0$ and $\theta = \pi$ are singular points of the differential equation (3.1.1). If we impose the condition that G is to be finite at the end points, then we find that the appropriate expression for G is given by

$$G\left(\theta, \theta'; \tau; \lambda\right) = \frac{\pi}{2} \frac{\Gamma\left(\frac{1}{2} + i\tau + \mu\right)}{\Gamma\left(\frac{1}{2} + i\tau - \mu\right)} \frac{P^{-\mu}_{-\frac{1}{2}+i\tau}(\cos \theta) \, P^{-\mu}_{-\frac{1}{2}+i\tau}(-\cos \theta')}{\cos[\pi(i\tau - \mu)]}, \quad (0 < \theta < \theta' < \pi),$$

$$(3.1.3)$$

where $\mu = \sqrt{\lambda}$, $\mathrm{Re} \sqrt{\lambda} > 0$. For $0 < \theta' < \theta < \pi$, the Green's function can be obtained from (3.1.3) by interchanging θ and θ'. It is obvious that G has only a branch point singularity at $\lambda = 0$ in the complex λ-plane and so a branch cut is chosen along the negative real axis of the λ-plane. As there are no other singularities, Γ can be taken as a circle at infinity not crossing the branch in the complex λ-plane. By Cauchy's theorem, the integral over the circle is equal to the integral over a contour which starts at infinity, runs parallel to but just above the negative real axis up to a point near to the point right of the origin, then runs parallel to but just below the negative real axis up to infinity. In the complex μ-plane, this contour transforms to a line parallel to the imaginary axis and without any loss of generality can be chosen as the imaginary axis itself. Thus, we obtain

$$\delta(\theta - \theta') \sin \theta' = \frac{1}{2i} \int_{-i\infty}^{i\infty} \mu \frac{\Gamma\left(\frac{1}{2} + i\tau + \mu\right)}{\Gamma\left(\frac{1}{2} + i\tau - \mu\right)} \frac{P_{-\frac{1}{2}+i\tau}^{-\mu}(\cos \theta)\, P_{-\frac{1}{2}+i\tau}^{-\mu}(-\cos \theta')}{\cos[\pi(i\tau - \mu)]} \, d\mu. \quad (3.1.4)$$

It may be interesting to show directly that the right side of (3.1.4) is actually equal to the left side. For this we note that, by using the estimates (1.2.45), as $|\mu| \to \infty$ in $\mathrm{Re}\, \mu > 0$

$$G \sim \frac{1}{2\mu} e^{\mu\alpha},$$

where $\alpha = \ln\left(\dfrac{\tan\frac{\theta}{2}}{\tan\frac{\theta'}{2}}\right)$ for $\theta < \theta'$. θ and θ' are to be interchanged for $\theta > \theta'$. Thus α is always negative. Hence, G vanishes exponentially as $|\mu| \to \infty$ in $\mathrm{Re}\, \mu > 0$ so that the integral is zero for $\theta \neq \theta'$. To examine the contribution when $\theta \to \theta'$, we let $\theta' = \theta + \epsilon$ where $\epsilon \to 0+$ and $\theta < \theta'$. Now

$$\tan\frac{\theta}{2} = \tan\frac{\theta'}{2}(1 - \epsilon \, \mathrm{cosec}\, \theta') + O(\epsilon^2)$$

so that $\alpha = -\epsilon \, \mathrm{cosec}\, \theta' + O(\epsilon^2)$.

Thus, the right side of (3.1.4) can be written as

$$\frac{1}{2\pi i} \int_{-i\infty}^{i\infty} e^{-\epsilon \mu \,\mathrm{cosec}\, \theta'} d\mu = \frac{1}{2\pi} \sin \theta' \int_{-\infty}^{\infty} e^{-i\epsilon t} dt = \sin \theta' \, \delta(\theta - \theta').$$

Now, the integral expansion formula for a function $f(\theta)$ defined on $(0, \pi)$ is formally obtained by using (1.5) as

$$f(\theta) = \frac{1}{2i} \int_{-i\infty}^{i\infty} \mu \frac{\Gamma\left(\frac{1}{2} + i\tau + \mu\right)}{\Gamma\left(\frac{1}{2} + i\tau - \mu\right)} \frac{P_{-\frac{1}{2}+i\tau}^{-\mu}(\cos \theta)}{\cos[\pi(i\tau - \mu)]} \left\{ \int_0^{\pi} \frac{f(\theta')}{\sin \theta'} P_{-\frac{1}{2}+i\tau}^{-\mu}(-\cos \theta') \, d\theta' \right\} d\mu.$$

$$(3.1.5)$$

Case II. $0 \le \theta \le \theta_2 < \pi$

(i) Let G be finite at $\theta = 0$ and $\dfrac{dG}{d\theta} = 0$ at $\theta = \theta_2$. Then for $\theta \le \theta'$,

$$G(\theta, \theta'; \tau; \lambda) = \frac{\pi}{2} \frac{\Gamma\left(\frac{1}{2} + i\tau + \mu\right)}{\Gamma\left(\frac{1}{2} + i\tau - \mu\right)} P_{-\frac{1}{2}+i\tau}^{-\mu}(\cos \theta) \frac{1}{\cos[\pi(i\tau - \mu)]} \times$$

$$\times \left[P_{-\frac{1}{2}+i\tau}^{-\mu}(-\cos \theta') - \frac{(\partial/\partial\theta_2) P_{-\frac{1}{2}+i\tau}^{-\mu}(-\cos \theta_2)}{(\partial/\partial\theta_2) P_{-\frac{1}{2}+i\tau}^{-\mu}(\cos \theta_2)} P_{-\frac{1}{2}+i\tau}^{-\mu}(\cos \theta') \right],$$

where $\mu = \sqrt{\lambda}$, $\mathrm{Re}\sqrt{\lambda} > 0$ and $(\partial/\partial\theta_2)\mathrm{P}^{-\mu}_{-\frac{1}{2}+i\tau}(\cos\theta_2) \equiv \left[(\partial/\partial\theta)\mathrm{P}^{-\mu}_{-\frac{1}{2}+i\tau}(\cos\theta)\right]_{\theta=\theta_2}$
and θ and θ' are to be interchanged when $\theta \geq \theta'$. As before, the Green's function G
has a branch point singularity at $\lambda = 0$ and as such a branch cut is chosen along the
negative real axis in the complex λ-plane. Proceeding as in Case I, we obtain

$$\delta(\theta - \theta')\sin\theta' = \frac{1}{2i}\int_{-i\infty}^{i\infty} \mu\, \frac{\Gamma\left(\frac{1}{2}+i\tau+\mu\right)\mathrm{P}^{-\mu}_{-\frac{1}{2}+i\tau}(\cos\theta)}{\Gamma\left(\frac{1}{2}+i\tau-\mu\right)\cos[\pi(i\tau-\mu)]} \times$$

$$\times \left[\mathrm{P}^{-\mu}_{-\frac{1}{2}+i\tau}(-\cos\theta') - \frac{(\partial/\partial\theta_2)\mathrm{P}^{-\mu}_{-\frac{1}{2}+i\tau}(-\cos\theta_2)}{(\partial/\partial\theta_2)\mathrm{P}^{-\mu}_{-\frac{1}{2}+i\tau}(\cos\theta_2)}\mathrm{P}^{-\mu}_{-\frac{1}{2}+i\tau}(\cos\theta')\right] d\mu.$$

The corresponding integral expansion for a function $f(\theta)$ in $(0, \theta_2)$ $(\theta_2 < \pi)$ is formally obtained as

$$f(\theta) = \frac{1}{2i}\int_{-i\infty}^{i\infty} \mu\, \frac{\Gamma\left(\frac{1}{2}+i\tau+\mu\right)}{\Gamma\left(\frac{1}{2}+i\tau-\mu\right)}\frac{\mathrm{P}^{-\mu}_{-\frac{1}{2}+i\tau}(\cos\theta)}{\cos[\pi(i\tau-\mu)]} \times$$

$$\times \left\{\int_0^{\theta_2}\frac{f(\theta')}{\sin\theta'}\left[\mathrm{P}^{-\mu}_{-\frac{1}{2}+i\tau}(-\cos\theta') - \frac{(\partial/\partial\theta_2)\mathrm{P}^{-\mu}_{-\frac{1}{2}+i\tau}(-\cos\theta_2)}{(\partial/\partial\theta_2)\mathrm{P}^{-\mu}_{-\frac{1}{2}+i\tau}(\cos\theta_2)}\mathrm{P}^{-\mu}_{-\frac{1}{2}+i\tau}(\cos\theta')\right] d\theta'\right\} d\mu.$$

$$(3.1.6)$$

(ii) Let G be finite at $\theta = 0$ and G $= 0$ at $\theta = \theta_2$.

Then the δ-function representation can be deduced as in relation (3.1.6) except
that the operator $(\partial/\partial\theta_2)$ is replaced by unity. The corresponding integral expansion
formula is obvious.

Case III. $0 < \theta_1 \leq \theta \leq \theta_2 < \pi$

(i) Let $\dfrac{dG}{d\theta} = 0$ at $\theta = \theta_1, \theta_2$. Then for $\theta \leq \theta'$,

$$G(\theta, \theta'; \tau; \lambda) = \frac{\pi}{2}\frac{\mathrm{M}(\theta, \theta_1; \tau; \lambda)\,\mathrm{M}(\theta', \theta_2; \tau; \lambda)}{\sin\pi\mu\,(\partial/\partial\theta_2)\mathrm{M}(\theta_2, \theta_1; \tau; \lambda)}, \qquad (3.1.7)$$

where

$$\mathrm{M}(\phi, \psi; \tau; \lambda) = \mathrm{P}^{\mu}_{-\frac{1}{2}+i\tau}(\cos\phi)\frac{\partial}{\partial\psi}\mathrm{P}^{-\mu}_{-\frac{1}{2}+i\tau}(\cos\psi) - \mathrm{P}^{-\mu}_{-\frac{1}{2}+i\tau}(\cos\phi)\frac{\partial}{\partial\psi}\mathrm{P}^{\mu}_{-\frac{1}{2}+i\tau}(\cos\psi).$$

The Green's function (3.1.7) has simple poles at $\lambda = \lambda_j$ $(j = 0, 1, 2, \ldots)$ along the
negative real λ-axis, where

$$\frac{\partial}{\partial\theta_2}\mathrm{M}(\theta_2, \theta_1; \tau; \lambda_j) = 0, \quad \lambda_j = -k_j^2 < 0.$$

In this case, the δ-function representation can be found as

$$\delta(\theta - \theta')\sin\theta' = \pi\sum_j \frac{k_j}{\sinh k_j\pi} \frac{(\partial/\partial\theta_2)\mathrm{P}^{ik_j}_{-\frac{1}{2}+i\tau}(\cos\theta_2)}{(\partial/\partial\theta_1)\mathrm{P}^{ik_j}_{-\frac{1}{2}+i\tau}(\cos\theta_1)} \times$$

$$\times \frac{\mathrm{M}(\theta,\theta_1;\tau;-k_j^2)\ \mathrm{M}(\theta',\theta_1;\tau;-k_j^2)}{(\partial^2/\partial(-k_j)\ \partial\theta_2)\mathrm{M}(\theta_2,\theta_1;\tau;-k_j^2)}, \tag{3.1.8}$$

for $\theta \leq \theta'$. In this case, the expansion of a function $f(\theta)$ defined on (θ_1,θ_2) is formally obtained as

$$f(\theta) = \pi\sum_j \frac{k_j}{\sinh k_j\pi} \frac{(\partial/\partial\theta_2)\mathrm{P}^{ik_j}_{-\frac{1}{2}+i\tau}(\cos\theta_2)\ \mathrm{M}(\theta,\theta_1;\tau,-k_j^2)}{(\partial^2/\partial(-k_j)\ \partial\theta_2)\mathrm{M}(\theta_2,\theta_1;\tau;-k_j^2)} \times$$

$$\times \left\{\int_{\theta_1}^{\theta_2} \frac{f(\theta')}{\sin\theta'}\ \mathrm{M}(\theta',\theta_1;\tau;-k_j^2)\ d\theta'\right\}. \tag{3.1.9}$$

(ii) Let $G = 0$ at $\theta = \theta_1,\theta_2$.

The expression for G and the δ-function representation in this case are the same as in relations (3.1.7) and (3.1.8) respectively, except that the operators $(\partial/\partial\theta_1)$ and $(\partial/\partial\theta_2)$ are replaced by unity. Also, the corresponding expansion of a function $f(\theta)$ defined on (θ_1,θ_2) can be formally obtained.

It may be noted that the integral expansion formulae given by (3.1.5), (3.1.6) and (3.1.9), etc., can be established rigorously under certain conditions (for each expansion formula) to be satisfied by the function $f(\theta)$. However, it is not trivial to obtain these conditions. They can be obtained by studying appropriate properties of the associated Legendre functions together with convergence criteria for the various integrals involving $f(\theta)$. We will see later how these conditions arise in a natural manner while establishing rigorously some integral expansions of somewhat similar type. However, here we assume that $f(\theta)$ satisfies the necessary conditions required for convergence of the various integrals.

3.1.2 Some integral expansion formulae can also be formally developed from the solution of some appropriate boundary value problems. Here we show how formula (3.1.5) can be obtained from the solution of Laplace's equation involving spherical coordinates in a suitable region. This has been established by Mandal (1971a) while considering the solution of Laplace's equation $\nabla^2 u = 0$ in the region $a \leq r < \infty$, $0 \leq \theta \leq \pi$ and $0 \leq \phi \leq \alpha$. Laplace's equation reduces to the form

$$\nu\,(\nu+1)\,v + \frac{1}{\sin\theta}\frac{\partial}{\partial\theta}\left(\sin\theta\,\frac{\partial v}{\partial\theta}\right) + \frac{1}{\sin^2\theta}\frac{\partial^2 v}{\partial\phi^2} = 0, \tag{3.1.10}$$

by the substitution $u(r,\theta,\phi) = r^{-\nu-1}\,v(\theta,\phi)$, Re $\nu > -1$ with $u \to 0$ as $r \to \infty$. The boundary conditions to be satisfied by $v(\theta,\phi)$ are taken as

$$\begin{aligned} v &= f(\theta), \quad \text{on } \phi = 0,\ 0 \leq \theta \leq \pi, \\ v &= 0, \quad\quad \text{on } \phi = \alpha,\ 0 \leq \theta \leq \pi, \end{aligned} \tag{3.1.11}$$

where $f(\theta)$ is some prescribed function.

Let

$$v_n = \int_0^\alpha v \sin\lambda\phi\,d\phi, \quad \lambda = \frac{n\pi}{\alpha},$$

then

$$v = \frac{2}{\alpha}\sum_{n=1}^{\infty} v_n(\theta)\sin\lambda\phi.$$

From the boundary conditions (3.1.11), it is observed that $v_n(\theta)$ satisfies the differential equation

$$\frac{d}{d\theta}\left(\sin\theta\,\frac{d\psi}{d\theta}\right) + \left\{\nu\,(\nu+1)\,\sin\theta - \frac{\lambda^2}{\sin\theta}\right\}\psi = -\frac{\lambda\,f(\theta)}{\sin\theta}. \tag{3.1.12}$$

The solution of this differential equation is

$$\psi(\theta) = -\lambda\int_0^\pi \frac{f(\theta')}{\sin\theta'}\,G(\theta,\theta';\nu;\lambda)\,d\theta', \tag{3.1.13}$$

where $G(\theta,\theta';\nu;\lambda)$ is the Green's function corresponding to the differential equation (3.1.12) and is given by

$$G(\theta,\theta';\nu;\lambda) = \frac{\pi}{2}\frac{\Gamma(1+\nu+\lambda)}{\Gamma(1+\nu-\lambda)}\frac{P_\nu^{-\lambda}(\cos\theta)\;P_\nu^{-\lambda}(-\cos\theta')}{\sin[\pi(1+\nu-\lambda)]}, \tag{3.1.14}$$

for $\theta < \theta'$. Interchanging θ and θ' in the relation (3.1.14), the corresponding Green's function can be obtained for $\theta' < \theta$.

Now G can be written as

$$G(\theta, \theta'; \nu; \lambda) = \frac{1}{2i} \int_L \mu \frac{\Gamma(1+\nu+\mu)}{\Gamma(1+\nu-\mu)} \frac{P_\nu^{-\mu}(\cos\theta) \, P_\nu^{-\mu}(-\cos\theta')}{\sin[\pi(1+\nu-\mu)]} \frac{d\mu}{(\mu^2 - \lambda^2)}, \quad (3.1.15)$$

where L is the straight line Re $\mu = c, -\lambda < \text{Re } \nu < c < \lambda$. The integrand has singularities only at $\mu = \pm\lambda$. As Re $\mu \to \infty$, the integrand behaves as $\frac{1}{2\mu} e^{\mu\alpha}$, where $\alpha = \ln\left(\frac{\tan\frac{\theta}{2}}{\tan\frac{\theta'}{2}}\right)$. For $\theta < \theta', \alpha$ is negative, so that the integrand vanishes as $|\mu| \to \infty$ in Re $\mu > 0$. For $\theta > \theta', \alpha = \ln\left(\frac{\tan\frac{\theta'}{2}}{\tan\frac{\theta}{2}}\right)$ so that α is again negative. Hence, for both cases the integrand in the relation (3.1.15) tends to zero as $|\mu| \to \infty$ in Re $\mu > 0$.

Taking a large semicircle in the half-plane Re $\mu \geq 0$ with L as its diameter, the only singularity of the integrand is at $\mu = \lambda$, and we observe that the integral in (3.1.15) reduces to (3.1.14). Thus from the representations (3.1.13) and (3.1.14), we obtain

$$v_n(\theta) = -\frac{\lambda}{2i} \int_L \mu \frac{\Gamma(1+\nu+\mu)}{\Gamma(1+\nu-\mu)} \frac{P_\nu^{-\mu}(\cos\theta) \, F(\mu)}{\sin[\pi(1+\nu-\mu)]} \frac{d\mu}{(\mu^2 - \lambda^2)}, \quad (3.1.16)$$

where

$$F(\mu) = \int_0^\pi \frac{f(\theta)}{\sin\theta} P_\nu^{-\mu}(-\cos\theta) \, d\theta. \quad (3.1.17)$$

Hence, we obtain

$$v(\theta, \phi) = \frac{2}{\alpha} \sum_{n=1}^\infty v_n \sin\lambda\phi. \quad (3.1.18)$$

Using the relation

$$\frac{\sin\mu(\alpha - \phi)}{\sin\mu\alpha} = -\frac{2}{\alpha} \sum_{n=1}^\infty \frac{\lambda \sin\lambda\phi}{(\mu^2 - \lambda^2)}, \quad (3.1.19)$$

equation (3.1.18) produces

$$v(\theta, \phi) = \frac{1}{2i} \int_L \mu \frac{\Gamma(1+\nu+\mu)}{\Gamma(1+\nu-\mu)} \frac{P_\nu^{-\mu}(\cos\theta)}{\sin[\pi(1+\nu-\mu)]} \frac{\sin\mu(\alpha - \phi)}{\sin\mu\alpha} F(\mu) \, d\mu. \quad (3.1.20)$$

Using the boundary condition on $\phi = 0$ in the relation (3.1.20), we find that

$$f(\theta) = \frac{1}{2\pi i} \int_L \mu \, \Gamma(\mu - \nu) \, \Gamma(1+\nu+\mu) \, P_\nu^{-\mu}(\cos\theta) \, F(\mu) \, d\mu, \quad (3.1.21)$$

71

where L is the line Re $\mu = c, -\lambda < \text{Re } \nu < c < \lambda$. As Re $\nu > -1$, and $\lambda = \frac{n\pi}{\alpha}$, making $c \to 0$, (3.1.21) becomes

$$f(\theta) = \frac{1}{2\pi i} \int_{-i\infty}^{i\infty} \mu \, \Gamma(\mu - \nu) \, \Gamma(1 + \nu + \mu) \, P_{\nu}^{-\mu}(\cos \theta) \, F(\mu) \, d\mu, \qquad (3.1.22)$$

where

$$F(\mu) = \int_0^{\pi} \frac{f(\theta)}{\sin \theta} \, P_{\nu}^{-\mu}(-\cos \theta) \, d\theta. \qquad (3.1.23)$$

Equation (3.1.23) gives the integral transform formula while equation (3.1.22) gives the corresponding inverse transform formula. The relations (3.1.22) and (3.1.23) together produce an integral expansion formula related to a Mehler–Fock type transform. For $\nu = -\frac{1}{2} + i\tau$, this is the same as (3.1.5) obtained earlier.

We can obtain the Parseval relation in connection with the integral transform defined by (3.1.23) as follows. Let $F(\mu)$ and $G(\mu)$ be the integral transforms of the type (3.1.23) of $f(\theta)$ and $g(\theta)$ respectively. Then

$$\int_{-i\infty}^{i\infty} \mu \, w(\mu) \, F(\mu) \, G(\mu) \, d\mu = \int_{-i\infty}^{i\infty} \mu \, w(\mu) \, F(\mu) \left\{ \int_0^{\pi} \frac{g(\pi - \theta)}{\sin \theta} \, P_{\nu}^{-\mu}(\cos \theta) \, d\theta \right\} d\mu, \qquad (3.1.24)$$

where $w(\mu)$ is to be determined. Assuming the interchange of the order of integration to be possible, the relation (3.1.24) becomes

$$\int_{-i\infty}^{i\infty} \mu \, w(\mu) \, F(\mu) \, G(\mu) \, d\mu = \int_0^{\pi} \frac{g(\pi - \theta)}{\sin \theta} \left\{ \int_{-i\infty}^{i\infty} \mu \, w(\mu) \, P_{\nu}^{-\mu}(\cos \theta) \, F(\mu) \, d\mu \right\} d\theta. \qquad (3.1.25)$$

By virtue of the inversion formula (3.1.22), $w(\mu)$ can be taken as

$$w(\mu) = \frac{\Gamma(1 + \nu + \mu) \, \Gamma(\mu - \nu)}{2\pi i}.$$

Hence, the Parseval relation becomes

$$\frac{1}{2\pi i} \int_{-i\infty}^{i\infty} \mu \, \Gamma(\mu - \nu) \, \Gamma(1 + \nu + \mu) \, F(\mu) \, G(\mu) \, d\mu$$

$$= \int_0^{\pi} \frac{1}{\sin \theta} \, f(\theta) \, g(\pi - \theta) \, d\theta = \int_0^{\pi} \frac{1}{\sin \theta} \, f(\pi - \theta) \, g(\theta) \, d\theta. \qquad (3.1.26)$$

This relation can be utilized to evaluate some particular integrals.

Example 3.1.1

If $f(\theta) = \sin^m \theta$, then

$$F(\mu) = \frac{\pi \, 2^{-\mu} \, \Gamma(\frac{m}{2} - \frac{\mu}{2}) \, \Gamma(\frac{m}{2} + \frac{\mu}{2})}{\Gamma(1 + \frac{m}{2} + \frac{\nu}{2}) \, \Gamma(\frac{m}{2} - \frac{\nu}{2}) \, \Gamma(1 + \frac{\nu}{2} + \frac{\mu}{2}) \, \Gamma(\frac{1}{2} - \frac{\nu}{2} + \frac{\mu}{2})}.$$

If $g(\theta) = \sin^n \theta$, then

$$G(\mu) = \frac{\pi \, 2^{-\mu} \, \Gamma(\frac{n}{2} - \frac{\mu}{2}) \, \Gamma(\frac{n}{2} + \frac{\mu}{2})}{\Gamma(1 + \frac{n}{2} + \frac{\nu}{2}) \, \Gamma(\frac{n}{2} - \frac{\nu}{2}) \, \Gamma(1 + \frac{\nu}{2} + \frac{\mu}{2}) \, \Gamma(\frac{1}{2} - \frac{\nu}{2} + \frac{\mu}{2})}.$$

Therefore, by using Parseval's relation (3.1.26),

$$I \equiv \frac{1}{2\pi i} \int_{-i\infty}^{i\infty} \mu \, \Gamma(1 + \nu + \mu) \, \Gamma(\mu - \nu) \, F(\mu) \, G(\mu) \, d\mu$$

$$= \int_0^\pi \sin^{m+n} \theta \, d\theta \quad (m + n > 0)$$

$$= 2^{m+n-1} \, \mathrm{B}(\frac{m+n}{2}, \frac{m+n}{2}). \tag{3.1.27}$$

Example 3.1.2

Following Saxena (1961), it is known that

$$\int_0^\pi \frac{P_\nu^{-\mu}(-\cos\theta)}{\sin\theta} \, G_{r,1}^{p,q} \left(x \sin^{2n} \theta \, \middle| \, \begin{matrix} \alpha_1, \alpha_2, \ldots, \alpha_r \\ \beta_1, \beta_2, \ldots, \beta_l \end{matrix} \right) d\theta$$

$$= \frac{\pi \, 2^{-\mu} \, n^{-\frac{1}{2}}}{\Gamma(1 + \frac{\nu}{2} + \frac{\mu}{2}) \, \Gamma(\frac{1}{2} - \frac{\nu}{2} + \frac{\mu}{2})} \, G_{r+2n,l+2n}^{p,q+2n} \left(x \, \middle| \, \begin{matrix} \delta_1, \delta_2, \ldots, \delta_{2n}, \alpha_1, \ldots, \alpha_r \\ \beta_1, \beta_2, \ldots, \beta_l, \gamma_1, \ldots, \gamma_{2n} \end{matrix} \right), \tag{3.1.28}$$

with

$$\delta_{k+1} = \frac{2 - \mu + 2k}{2n}, \quad \delta_{n+k+1} = \frac{2 + \mu + 2k}{2n}, \quad \gamma_{k+1} = \frac{1 - \nu + 2k}{2n}, \quad \gamma_{n+k+1} = \frac{2 + \nu + 2k}{2n},$$

$k = 0, 1, \ldots, (n-1)$ and n is a positive integer, under the conditions

(i) $0 \le p \le 1$, $0 \le q \le r < 1$ (or $r = 1$ and $|x| < 1$),

$$\mathrm{Re} \, (\pm\mu + 2n\beta_j) > 0, \quad (j = 1, 2, \ldots, p);$$

(ii) $0 \le q \le r, 1 \le p \le l < r, 2(p+q) > r + l, |\arg x| < \pi(p + q - \frac{r}{2} - \frac{1}{2})$,

$$\mathrm{Re} \, (\pm\mu + 2n\beta_j) > 0, \quad (j = 1, 2, \ldots, p);$$

73

(iii) $0 \leq q \leq r, 0 \leq p \leq l \leq r, 2q > r + l, |\arg x| < \pi(q - \frac{r}{2} - \frac{\nu}{2})$,

$$\text{Re}\,(\pm\mu + 2n\beta_j) > 0, \quad (j = 1, 2, \ldots, p),$$

where G is Meijer's G-function. Hence taking

$$f(\theta) = \sin^m \theta \text{ and } g(\theta) = G_{r,l}^{p,q}\left(x\sin^{2n}\theta \,\middle|\, \begin{matrix} \alpha_1, \ldots, \alpha_r \\ \beta_1, \ldots, \beta_l \end{matrix}\right),$$

with the help of Parseval's relation (3.1.26), we obtain

$$I_2 \equiv \frac{1}{2\pi i} \int_{-i\infty}^{i\infty} \frac{\pi^2 \, 2^{-2\mu} \, \Gamma(\mu - \nu) \, \Gamma(1 + \nu + \mu) \, \Gamma(\frac{m}{2} + \frac{\mu}{2}) \, \Gamma(\frac{m}{2} - \frac{\mu}{2})}{\sqrt{n}\{\Gamma(\frac{1}{2} - \frac{\nu}{2} + \frac{\mu}{2}) \, \Gamma(1 + \frac{\nu}{2} + \frac{\mu}{2})\}^2 \, \Gamma(\frac{1}{2} + \frac{m}{2} + \frac{\nu}{2}) \, \Gamma(\frac{m}{2} - \frac{\nu}{2})} \times$$

$$\times G_{r+2n,l+2n}^{p,q+2n}\left(x \,\middle|\, \begin{matrix} \delta_1, \ldots, \delta_{2n}, \alpha_1, \ldots, \alpha_r \\ \beta_1, \ldots, \beta_l, \gamma_1, \ldots, \gamma_{2n} \end{matrix}\right) \mu \, d\mu$$

$$= \int_0^\pi \sin^{m-1}\theta \, G_{r,l}^{p,q}\left(x\sin^{2n}\theta \,\middle|\, \begin{matrix} \alpha_1, \alpha_2, \ldots, \alpha_r \\ \beta_1, \beta_2, \ldots, \beta_l \end{matrix}\right) d\theta. \tag{3.1.29}$$

From the definition of Meijer's G-function, by changing the order of integration (which is assumed to be permissible), we obtain

$$I_2 = \frac{1}{2\pi i} \int_C \frac{\prod\limits_{j=1}^{p}\Gamma(\beta_j - \xi)\prod\limits_{j=1}^{q}(1 - \alpha_j + \xi)}{\prod\limits_{j=p+1}^{l}\Gamma(1 - \beta_j + \xi)\prod\limits_{j=q+1}^{r}\Gamma(\alpha_j - \xi)} \, x^\xi\left\{\int_0^\pi \sin^{2n\xi + m - 1}\theta \, d\theta\right\} d\xi, \tag{3.1.30}$$

where C is a loop starting and ending at $+\infty$ and enclosing all the poles of $\Gamma(\beta_j - \xi)$ $(j = 1, 2, \ldots, p)$, once in the negative direction, but none of the poles of $\Gamma(1 - \alpha_j + \xi) = 0$ $(j = 1, 2, \ldots, q)$, and $\Gamma(\xi + \frac{m-1+2k}{2n}) = 0$ $(k = 0, 1, \ldots, n - 1)$. Then the integral (3.1.30) becomes

$$I_2 = \frac{1}{2\pi i} \int_C \frac{2\pi^{\frac{1}{2}}}{n^{\frac{1}{2}}} \frac{\prod\limits_{j=1}^{p}\Gamma(\beta_j - \xi)\prod\limits_{j=1}^{q}(1 - \alpha_j + \xi)\prod\limits_{k=0}^{n-1}\Gamma(\xi + \frac{m-1+2k}{2n})}{\prod\limits_{j=p+1}^{l}\Gamma(1 - \beta_j + \xi)\prod\limits_{j=q+1}^{r}\Gamma(\alpha_j - \xi)\prod\limits_{k=0}^{n-1}\Gamma(\xi + \frac{n\lambda+2k}{2n})} \, x^\xi \, d\xi$$

$$= \frac{2\pi^{\frac{1}{2}}}{n^{\frac{1}{2}}} G_{r+n,l+n}^{p,q+n}\left(x \,\middle|\, \begin{matrix} \tau_1, \ldots, \tau_n, \alpha_1, \ldots, \alpha_r \\ \beta_1, \ldots, \beta_l, \theta_1 \ldots, \theta_n \end{matrix}\right), \tag{3.1.31}$$

where $\tau_{k+1} = \frac{1-m+2k}{2n}$, $\theta_{k+1} = \frac{2k-m}{2n}$.

3.1.3 Lebedev and Skal'skaya (1986) established rigorously two integral expansion formulae which are somewhat similar to the integral expansions given by (3.1.5) and (3.1.22). Their main result is stated in the form of the following theorem.

Theorem 3.1.1

Let $f(x)$ be an arbitrary function defined on $(-1,1)$ and satisfying the following conditions:

(i) $f(x)$ is piecewise continuous and has bounded variation in $(-1,1)$,

(ii) $f(x)(1-x^2)^{-1}\ln(1-x^2)^{-1} \in L(-1,1)$.

Then

$$\frac{1}{2}[f(x+0)+f(x-0)] = \frac{1}{2\pi i}\int_{-\infty}^{\infty} \sigma\,\Gamma(\frac{1}{2}+i\tau-i\sigma)\,\Gamma(\frac{1}{2}-i\tau-i\sigma)F_1(\sigma)\,P^{i\sigma}_{-\frac{1}{2}+i\tau}(x)\,d\sigma,$$
(3.1.32)

with $F_1(\sigma) = \int_{-1}^{1}\frac{f(x)}{(1-x^2)}\,P^{i\sigma}_{-\frac{1}{2}+i\tau}(-x)\,dx;$ and

$$\frac{1}{2}[f(x+0)+f(x-0)] = \frac{1}{2\pi i}\int_{-\infty}^{\infty} \sigma\,\Gamma(\frac{1}{2}+i\tau-i\sigma)\,\Gamma(\frac{1}{2}-i\tau-i\sigma)F_2(\sigma)\,P^{i\sigma}_{-\frac{1}{2}+i\tau}(-x)\,d\sigma,$$
(3.1.33)

with $F_2(\sigma) = \int_{-1}^{1}\frac{f(x)}{(1-x^2)}\,P^{i\sigma}_{-\frac{1}{2}+i\tau}(x)\,dx,$
where τ is an arbitrary given real parameter.

Proof:

The theorem is proved utilizing the properties of the associated Legendre functions. Use of the inequality (1.2.24) produces

$$\int_{-1}^{1}\left|\frac{f(x)}{(1-x^2)}\,P^{i\sigma}_{-\frac{1}{2}+i\tau}(-x)\right|dx \leq \sqrt{\frac{\sinh\pi\sigma}{\pi\sigma}}\int_{-1}^{1}\frac{|f(x)|}{(1-x^2)}\,P_{-\frac{1}{2}+i\tau}(-x)\,dx.$$

This shows that the conditions imposed on $f(x)$ imply that the integral $F_1(\sigma)$ is absolutely and uniformly convergent for $\sigma \in [-T,T]$ $(T>0)$. Hence, $F_1(\sigma)$ is continuous on $[-T,T]$ and the repeated integral

$$J(x,T) = \frac{1}{2\pi i}\int_{-T}^{T}\sigma\,\Gamma(\frac{1}{2}+i\tau-i\sigma)\,\Gamma(\frac{1}{2}-i\tau-i\sigma)\,P^{i\sigma}_{-\frac{1}{2}+i\tau}(x)\times$$

$$\times\left\{\int_{-1}^{1}\frac{f(y)}{(1-y^2)}\,P^{i\sigma}_{-\frac{1}{2}+i\tau}(-y)\,dy\right\}d\sigma$$

75

has a meaning. Further, by virtue of the uniform convergence, changing the order of integration, we obtain

$$J(x,T) = \int_{-1}^{1} \frac{f(y)}{(1-y^2)} \, K(x,y,T) \, dy, \qquad (3.1.34)$$

where

$$K(x,y,T) = \frac{1}{2\pi i} \int_{-T}^{T} \sigma \, \Gamma(\tfrac{1}{2}+i\tau-i\sigma) \, \Gamma(\tfrac{1}{2}-i\tau-i\sigma) \, P^{i\sigma}_{-\frac{1}{2}+i\tau}(x) \, P^{i\sigma}_{-\frac{1}{2}+i\tau}(-y) \, d\sigma. \qquad (3.1.35)$$

Now we shall show that the kernel $K(x,y,T)$ is symmetric in the variables x and y. By definition, we have

$$K(x,y,T) - K(y,x,T) = \frac{1}{2\pi i} \int_{-T}^{T} \sigma \, \Gamma(\tfrac{1}{2}+i\tau-i\sigma) \, \Gamma(\tfrac{1}{2}-i\tau-i\sigma) \times$$

$$\times \left[P^{i\sigma}_{-\frac{1}{2}+i\tau}(x) \, P^{i\sigma}_{-\frac{1}{2}+i\tau}(-y) - P^{i\sigma}_{-\frac{1}{2}+i\tau}(-x) \, P^{i\sigma}_{-\frac{1}{2}+i\tau}(y) \right] \, d\sigma.$$

It follows from the properties of associated Legendre functions (cf. Erdélyi et al. 1953, p. 144) that the integrand in the above integral is an odd function of σ, hence the integral vanishes. Thus

$$K(x,y,T) = K(y,x,T). \qquad (3.1.36)$$

To investigate the behaviour of $K(x,y,T)$ as $T \to \infty$, by writing $\mu = -i\sigma$, we write (3.1.35) as

$$K(x,y,T) = \frac{1}{2\pi i} \int_{-iT}^{iT} \mu \, \Gamma(\tfrac{1}{2}+i\tau+\mu) \, \Gamma(\tfrac{1}{2}-i\tau+\mu) \, P^{-\mu}_{-\frac{1}{2}+i\tau}(x) \, P^{-\mu}_{-\frac{1}{2}+i\tau}(-y) \, d\mu. \quad (3.1.37)$$

The expression under the integral sign in equation (3.1.37) is a function of the complex variable μ, regular in the semi-plane $\text{Re } \mu \geq 0$. Therefore, the integration along a segment of the imaginary axis can be replaced by integration along a circular arc Γ_T of radius T in this semi-plane. Thus

$$K(x,y,T) = \frac{1}{2\pi i} \int_{\Gamma_T} \mu \, \Gamma(\tfrac{1}{2}+i\tau+\mu) \, \Gamma(\tfrac{1}{2}-i\tau+\mu) \, P^{-\mu}_{-\frac{1}{2}+i\tau}(x) \, P^{-\mu}_{-\frac{1}{2}+i\tau}(-y) \, d\mu. \quad (3.1.38)$$

Now we fix x and suppose that $y \leq x$. By virtue of the relations given in (1.2.44) and the asymptotic properties of the gamma function, we conclude that

$$\mu\, \Gamma(\tfrac{1}{2} + i\tau + \mu)\, \Gamma(\tfrac{1}{2} - i\tau + \mu)\, P^{-\mu}_{-\frac{1}{2}+i\tau}(x)\, P^{-\mu}_{-\frac{1}{2}+i\tau}(-y)$$

$$= \left(\frac{1+x}{1-x}\frac{1-y}{1+y}\right)^{-\mu/2}\left[1 + O(|\mu|^{-1})\right], \tag{3.1.39}$$

where $x \in (-1,1)$ is fixed, $y \in (-1,x]$ and $\mu \in \Gamma_T$.

Let $\xi = \tfrac{1}{2}\ln\dfrac{1+x}{1-x}, \eta = \tfrac{1}{2}\ln\dfrac{1+y}{1-y}$. Then from equations (3.1.37) $-$ (3.1.39), we obtain for $y \leq x$ (i.e., for $\eta \leq \xi$)

$$K(x,y,T) = \frac{1}{2\pi i}\int_{\Gamma_T} \exp\left\{-\mu(\xi - \eta)\right\} d\mu + O(1)\int_0^{\pi/2} exp\{-T(\xi - \eta)\cos\phi\}\, d\phi$$

$$= \frac{1}{\pi}\frac{\sin T(\xi - \eta)}{(\xi - \eta)} + O(1)\frac{1 - \exp\{-T(\xi - \eta)\}}{T(\xi - \eta)}. \tag{3.1.40}$$

For $y \geq x$, using the symmetry property (3.1.36) and the representation (3.1.38) of $K(x,y,T)$ with the variables x,y replaced by y,x as in formula (3.1.40), we obtain

$$K(x,y,T) = \frac{1}{\pi}\frac{\sin T(\eta - \xi)}{(\eta - \xi)} + O(1)\frac{1 - \exp\{-T(\eta - \xi)\}}{T(\eta - \xi)}. \tag{3.1.41}$$

Now we divide $J(x,T)$ into two parts as

$$J(x,T) = \int_{-1}^{x}\frac{f(y)}{1 - y^2}\, K(x,y,T)\, dy + \int_{x}^{1}\frac{f(y)}{1 - y^2}\, K(x,y,T)\, dy$$

$$= J_1(x,T) + J_2(x,T). \tag{3.1.42}$$

Then the relation (3.1.40) implies that

$$J_1(x,T) = \frac{1}{\pi}\int_{-\infty}^{\xi} f(\tanh\eta)\frac{\sin T(\xi - \eta)}{(\xi - \eta)}\, d\eta$$

$$+ O(1)\int_{-\infty}^{\xi}|f(\tanh\eta)|\frac{1 - \exp\{-T(\xi - \eta)\}}{T(\xi - \eta)}\, d\eta. \tag{3.1.43}$$

The conditions satisfied by $f(x)$ imply that $f(\tanh\eta) \in L(-\infty,\infty)$ and hence, as $T \to \infty$

$$\frac{1}{\pi}\int_{-\infty}^{\xi} f(\tanh\eta)\frac{\sin T(\xi - \eta)}{(\xi - \eta)}\, d\eta = \frac{1}{2}f(\tanh\xi - 0) + o(1)$$

$$= \frac{1}{2}f(x - 0) + o(1). \tag{3.1.44}$$

Writing the second integral in equation (3.1.43) as the sum of integrals over $(\xi - \delta, \xi)$ and $(-\infty, \xi - \delta)$ and then using the inequality

$$\frac{1 - \exp\{-T(\xi - \eta)\}}{T(\xi - \eta)} \leq \begin{cases} 1, & \xi - \delta \leq \eta \leq \xi, \\ \dfrac{1}{\delta T}, & -\infty < \eta \leq \xi - \delta, \end{cases} \qquad (3.1.45)$$

we find that as $T \to \infty$ (for a sufficiently small δ)

$$O(1) \int_{-\infty}^{\xi} |f(\tanh \eta)| \frac{1 - \exp\{-T(\xi - \eta)\}}{T(\xi - \eta)} \, d\eta$$

$$= O(1) \int_{\xi - \delta}^{\xi} |f(\tanh \eta)| \, d\eta + O(1) \frac{1}{\delta T} \int_{-\infty}^{\xi - \delta} |f(\tanh \eta)| \, d\eta$$

$$= o(1) + O(T^{-1}) = o(1). \qquad (3.1.46)$$

Hence, it follows from (3.1.44) and (3.1.46) that

$$\lim_{T \to \infty} J_1(x, T) = \frac{1}{2} f(x - 0). \qquad (3.1.47)$$

Using the relation (3.1.41) in $J_2(x, T)$ we obtain

$$\lim_{T \to \infty} J_2(x, T) = \frac{1}{2} f(x + 0). \qquad (3.1.48)$$

Thus

$$\lim_{T \to \infty} J(x, T) = \frac{1}{2} [f(x + 0) + f(x - 0)], \qquad (3.1.49)$$

and this proves the expansion formula (3.1.32). The proof of formula (3.1.33) can be established in a similar way.

Adding the formulae (3.1.32) and (3.1.33), we find that

$$\frac{1}{2} [f(x + 0) + f(x - 0)] = \frac{1}{4\pi i} \int_{-\infty}^{\infty} \sigma \, \Gamma(\frac{1}{2} + i\tau - i\sigma) \, \Gamma(\frac{1}{2} - i\tau - i\sigma) \times$$

$$\times [F_1(\sigma) \, P^{i\sigma}_{-\frac{1}{2} + i\tau}(x) + F_2(\sigma) \, P^{i\sigma}_{-\frac{1}{2} + i\tau}(-x)] \, d\sigma. \qquad (3.1.50)$$

The above integral expansion formula takes a simpler form when $f(x)$ is an even or an odd function. When $f(x)$ is even,

$$F_1(\sigma) = F_2(\sigma) = \int_0^1 \frac{f(x)}{1 - x^2} \left[P^{i\sigma}_{-\frac{1}{2} + i\tau}(x) + P^{i\sigma}_{-\frac{1}{2} + i\tau}(-x) \right] dx,$$

78

and when $f(x)$ is odd,

$$F_1(\sigma) = -F_2(\sigma) = \int_0^1 \frac{f(x)}{1-x^2} \left[P^{i\sigma}_{-\frac{1}{2}+i\tau}(x) - P^{i\sigma}_{-\frac{1}{2}+i\tau}(-x) \right] dx.$$

Then the following expansion formula can be obtained as

$$\frac{1}{2}[f(x+0) + f(x-0)] = \frac{1}{4\pi i} \int_{-\infty}^{\infty} \sigma\, \Gamma(\frac{1}{2} + i\tau - i\sigma)\, \Gamma(\frac{1}{2} - i\tau - i\sigma)\, F_1(\sigma) \times$$

$$\times \left[P^{i\sigma}_{-\frac{1}{2}+i\tau}(x) + P^{i\sigma}_{-\frac{1}{2}+i\tau}(-x) \right] d\sigma, \quad -1 < x < 1, \tag{3.1.51}$$

when $f(-x) = f(x)$ and

$$\frac{1}{2}[f(x+0) + f(x-0)] = \frac{1}{4\pi i} \int_{-\infty}^{\infty} \sigma\, \Gamma(\frac{1}{2} + i\tau - i\sigma)\, \Gamma(\frac{1}{2} - i\tau - i\sigma)\, F_1(\sigma) \times$$

$$\times \left[P^{i\sigma}_{-\frac{1}{2}+i\tau}(x) - P^{i\sigma}_{-\frac{1}{2}+i\tau}(-x) \right] d\sigma, \quad -1 < x < 1, \tag{3.1.52}$$

when $f(-x) = -f(x)$. Writing the integrals in equations (3.1.51) and (3.1.52) as sums of integrals over the intervals $(-\infty, 0)$ and $(0, \infty)$, then using the relations

$$\Gamma(\frac{1}{2} + i\tau - i\sigma)\Gamma(\frac{1}{2} - i\tau - i\sigma) \left[P^{i\sigma}_{-\frac{1}{2}+i\tau}(0) \right]^2 - \Gamma(\frac{1}{2} + i\tau + i\sigma)\Gamma(\frac{1}{2} - i\tau + i\sigma) \left[P^{-i\sigma}_{-\frac{1}{2}+i\tau}(0) \right]^2$$

$$= \frac{i}{2\pi^2} \left| \Gamma\left(\frac{1}{4} + \frac{i\tau}{2} + \frac{i\sigma}{2}\right) \Gamma\left(\frac{1}{4} + \frac{i\tau}{2} - \frac{i\sigma}{2}\right) \right|^2 \sinh \pi\sigma,$$

and

$$\Gamma(\frac{1}{2} + i\tau - i\sigma)\Gamma(\frac{1}{2} - i\tau - i\sigma) \left[P^{i\sigma}_{-\frac{1}{2}+i\tau}{}'(0) \right]^2 - \Gamma(\frac{1}{2} + i\tau + i\sigma)\Gamma(\frac{1}{2} - i\tau + i\sigma) \left[P^{-i\sigma}_{-\frac{1}{2}+i\tau}{}'(0) \right]^2$$

$$= \frac{2}{\pi^2 i} \left| \Gamma\left(\frac{3}{4} + \frac{i\tau}{2} + \frac{i\sigma}{2}\right) \Gamma\left(\frac{3}{4} + \frac{i\tau}{2} - \frac{i\sigma}{2}\right) \right|^2 \sinh \pi\sigma,$$

we obtain the following integral expansion formulae for $f(x)$ in the interval $(0, 1)$:

$$\frac{1}{2}[f(x+0) + f(x-0)] = \frac{1}{2\pi^3} \int_0^{\infty} \sigma\, \sinh \pi\sigma \left| \Gamma(\frac{1}{4} + \frac{i\tau}{2} + \frac{i\sigma}{2})\Gamma(\frac{1}{4} + \frac{i\tau}{2} - \frac{i\sigma}{2}) \right|^2 \times$$

$$\times \frac{P^{i\sigma}_{-\frac{1}{2}+i\tau}(x) + P^{i\sigma}_{-\frac{1}{2}+i\tau}(-x)}{2P^{i\sigma}_{-\frac{1}{2}+i\tau}(0)} \left\{ \int_0^1 \frac{f(y)}{1-y^2} \frac{P^{i\sigma}_{-\frac{1}{2}+i\tau}(y) + P^{i\sigma}_{-\frac{1}{2}+i\tau}(-y)}{2P^{i\sigma}_{-\frac{1}{2}+i\tau}(0)} dy \right\} d\sigma, \quad 0 < x < 1,$$

$$\tag{3.1.53}$$

and

$$\frac{1}{2}[f(x+0)+f(x-0)] = \frac{2}{\pi^3}\int_0^\infty \sigma\, \sinh\pi\sigma \left|\Gamma(\frac{3}{4}+\frac{i\tau}{2}+\frac{i\sigma}{2})\Gamma(\frac{3}{4}+\frac{i\tau}{2}-\frac{i\sigma}{2})\right|^2 \times$$

$$\times \frac{P^{i\sigma}_{-\frac{1}{2}+i\tau}(x)-P^{i\sigma}_{-\frac{1}{2}+i\tau}(-x)}{2P^{i\sigma}_{-\frac{1}{2}+i\tau}{}'(0)}\left\{\int_0^1 \frac{f(y)}{1-y^2}\frac{P^{i\sigma}_{-\frac{1}{2}+i\tau}(y)-P^{i\sigma}_{-\frac{1}{2}+i\tau}(-y)}{2P^{i\sigma}_{-\frac{1}{2}+i\tau}{}'(0)}dy\right\}d\sigma,\ 0<x<1.$$

$$(3.1.54)$$

The formulae (3.1.53) and (3.1.54) can be considered as canonical representations of the function $f(x)$ in the interval $(0,1)$. The combination of spherical harmonics in relations (3.1.53) and (3.1.54) can also be expressed in terms of hypergeometric functions as given by

$$\frac{P^{i\sigma}_{-\frac{1}{2}+i\tau}(x)+P^{i\sigma}_{-\frac{1}{2}+i\tau}(-x)}{2P^{i\sigma}_{-\frac{1}{2}+i\tau}(0)} = (1-x^2)^{-i\sigma/2}\ F\left(\frac{1}{4}+\frac{i\tau}{2}-\frac{i\sigma}{2},\frac{1}{4}-\frac{i\tau}{2}-\frac{i\sigma}{2};\frac{1}{2};x^2\right),$$

$$(3.1.55)$$

and

$$\frac{P^{i\sigma}_{-\frac{1}{2}+i\tau}(x)-P^{i\sigma}_{-\frac{1}{2}+i\tau}(-x)}{2P^{i\sigma}_{-\frac{1}{2}+i\tau}{}'(0)} = x(1-x^2)^{-i\sigma/2}F\left(\frac{3}{4}+\frac{i\tau}{2}-\frac{i\sigma}{2},\frac{3}{4}-\frac{i\tau}{2}-\frac{i\sigma}{2};\frac{3}{2};x^2\right).$$

$$(3.1.56)$$

Now we give some examples of the integral expansion formulae established above.

Example 3.1.3

Let

$$f(x) = \begin{cases} 0, & -1<x<0 \\ P^{-\mu}_{-\frac{1}{2}+i\tau}(x), & 0<x<1. \end{cases}$$

Then, from formula (3.1.32), we obtain

$$f(x) = \frac{1}{2\pi i}\int_{-\infty}^\infty \frac{2^{-\mu-i\sigma}}{(\mu^2+\sigma^2)}\ \sigma\ \left[\frac{\Gamma(\frac{1}{4}+\frac{i\tau}{2}-\frac{i\sigma}{2})\ \Gamma(\frac{1}{4}-\frac{i\tau}{2}-\frac{i\sigma}{2})}{\Gamma(\frac{1}{4}+\frac{i\tau}{2}+\frac{\mu}{2})\ \Gamma(\frac{1}{4}-\frac{i\tau}{2}+\frac{\mu}{2})}\right.$$

$$\left.+\frac{\Gamma(\frac{3}{4}+\frac{i\tau}{2}-\frac{i\sigma}{2})\ \Gamma(\frac{3}{4}-\frac{i\tau}{2}-\frac{i\sigma}{2})}{\Gamma(\frac{3}{4}+\frac{i\tau}{2}+\frac{\mu}{2})\ \Gamma(\frac{3}{4}-\frac{i\tau}{2}+\frac{\mu}{2})}\right] P^{i\sigma}_{-\frac{1}{2}+i\tau}(x)\ d\sigma,\qquad (3.1.57)$$

where $-1<x<1$, Re $\mu>0$.

Example 3.1.4

$$\sqrt{1-x^2} = \frac{\pi^{-\frac{3}{2}}}{2\,\Gamma(\frac{3}{4}+\frac{i\tau}{2})\,\Gamma(\frac{3}{4}-\frac{i\tau}{2})}\int_0^\infty \sigma\,\sinh\frac{\pi\sigma}{2}\left|\Gamma(\frac{1}{4}+\frac{i\tau}{2}+\frac{i\sigma}{2})\,\Gamma(\frac{1}{4}+\frac{i\tau}{2}-\frac{i\sigma}{2})\right|^2$$

$$\times\frac{P^{i\sigma}_{-\frac{1}{2}+i\tau}(x)+P^{i\sigma}_{-\frac{1}{2}+i\tau}(-x)}{2P^{i\sigma}_{-\frac{1}{2}+i\tau}(0)}\,d\sigma,\quad 0\le x<1. \tag{3.1.58}$$

This result can be obtained from the expansion formula (3.1.53).

Example 3.1.5

$$x\sqrt{1-x^2} = \frac{\pi^{-\frac{3}{2}}}{\Gamma(\frac{5}{4}+\frac{i\tau}{2})\,\Gamma(\frac{5}{4}-\frac{i\tau}{2})}\int_0^\infty \sigma\,\sinh\frac{\pi\sigma}{2}\left|\Gamma\left(\frac{3}{4}+\frac{i\tau}{2}+\frac{i\sigma}{2}\right)\Gamma\left(\frac{3}{4}+\frac{i\tau}{2}-\frac{i\sigma}{2}\right)\right|^2$$

$$\times\frac{P^{i\sigma}_{-\frac{1}{2}+i\tau}(x)-P^{i\sigma}_{-\frac{1}{2}+i\tau}(-x)}{2P^{i\sigma}_{-\frac{1}{2}+i\tau}{}'(0)}\,d\sigma,\quad 0\le x<1. \tag{3.1.59}$$

The above result is obtained by using formula (3.1.54).

Example 3.1.6

$$(1-x^2)^{\frac{\mu}{2}} = \frac{\pi^{-\frac{5}{2}}}{4\,\Gamma(\frac{1}{4}+\frac{\mu}{2}+\frac{i\tau}{2})\,\Gamma(\frac{1}{4}+\frac{\mu}{2}-\frac{i\tau}{2})}\int_0^\infty \sigma\,\sinh\pi\sigma\,\Gamma\left(\frac{\mu}{2}+\frac{i\sigma}{2}\right)\Gamma\left(\frac{\mu}{2}-\frac{i\sigma}{2}\right)$$

$$\times\left|\Gamma\left(\frac{1}{4}+\frac{i\tau}{2}+\frac{i\sigma}{2}\right)\Gamma\left(\frac{1}{4}+\frac{i\tau}{2}-\frac{i\sigma}{2}\right)\right|^2\frac{P^{i\sigma}_{-\frac{1}{2}+i\tau}(x)+P^{i\sigma}_{-\frac{1}{2}+i\tau}(-x)}{2P^{i\sigma}_{-\frac{1}{2}+i\tau}(0)}\,d\sigma,\quad 0\le x<1,$$

$$\text{Re}\,\mu>0. \tag{3.1.60}$$

Example 3.1.7

$$\frac{P^{-\mu}_{-\frac{1}{2}+i\tau}(\sqrt{1-x^2})}{x^\mu\,(1-x^2)^{-\frac{\mu}{2}}} = \frac{\pi^{-\frac{5}{2}}}{2^{1+\mu}\,\Gamma(\frac{1}{2}+\mu)\,\Gamma(\frac{1}{2}+\mu+i\tau)\,\Gamma(\frac{1}{2}+\mu-i\tau)}\int_0^\infty \sigma\,\sinh\pi\sigma$$

$$\times\Gamma(\mu+i\sigma)\Gamma(\mu-i\sigma)\left|\Gamma\left(\frac{1}{4}+\frac{i\tau}{2}+\frac{i\sigma}{2}\right)\Gamma\left(\frac{1}{4}+\frac{i\tau}{2}-\frac{i\sigma}{2}\right)\right|^2\times$$

$$\times\frac{P^{i\sigma}_{-\frac{1}{2}+i\tau}(x)+P^{i\sigma}_{-\frac{1}{2}+i\tau}(-x)}{2P^{i\sigma}_{-\frac{1}{2}+i\tau}(0)}\,d\sigma,\quad 0\le x<1,\ \text{Re}\,\mu>0. \tag{3.1.61}$$

In all the above results, the conditions under which the appropriate theorems hold are satisfied. Another interesting example is

81

Example 3.1.8

$$1 = \frac{\pi^{-\frac{3}{2}}}{\Gamma(\frac{1}{4} + \frac{i\tau}{2})\,\Gamma(\frac{1}{4} - \frac{i\tau}{2})} \int_0^\infty \cosh\frac{\pi\sigma}{2} \left| \Gamma\left(\frac{1}{4} + \frac{i\tau}{2} + \frac{i\sigma}{2}\right) \Gamma\left(\frac{1}{4} + \frac{i\tau}{2} - \frac{i\sigma}{2}\right) \right|^2 \times$$

$$\times \frac{P^{i\sigma}_{-\frac{1}{2}+i\tau}(x) + P^{i\sigma}_{-\frac{1}{2}+i\tau}(-x)}{2P^{i\sigma}_{-\frac{1}{2}+i\tau}(0)} \, d\sigma, \quad 0 \le x < 1. \tag{3.1.62}$$

This integral expansion cannot be obtained from formula (3.1.53) directly because the interior integral diverges. Formally, the formula (3.1.62) can be obtained from equation (3.1.60) by letting $\mu \to 0$. This formula can be proved directly by evaluating the integral on the right by the residue method.

3.1.4 Belichenko (1987a) generalized the expansion formula (3.1.22) for arbitrary complex ν and established rigorously the proof of the expansion formula by directly investigating the properties of associated spherical harmonics and thus giving the conditions which a function has to satisfy to be a candidate for such an expansion. His result is given in the form of the following theorem.

Theorem 3.1.2

Let $f(x)$ be a function defined on $(0, \pi)$ and satisfying the following conditions:

(i) $f(x)$ is piecewise continuous and possesses bounded variation in $(0, \pi)$,

(ii) $\dfrac{f(x)}{\sin x} \ln \sin \dfrac{x}{2} \in L(0, a)$, and $\dfrac{f(x)}{\sin x} \in L(a, \pi)$, $0 < a < \pi$.

Then

$$\tfrac{1}{2}[f(x+0) + f(x-0)] = -\sum_{m=0}^{[\text{Re }\nu - \frac{1}{2}]} (\tfrac{1}{2} + m - \nu) \frac{\Gamma(2\nu - m)}{m!} P^{\frac{1}{2}+m-\nu}_{\nu-\frac{1}{2}}(\cos x) \times$$

$$\times \int_0^\pi \frac{f(y)}{\sin y} P^{\frac{1}{2}+m-\nu}_{\nu-\frac{1}{2}}(\cos y) \, dy + \frac{1}{2\pi i} \int_{-i\infty}^{i\infty} \mu\, \Gamma(\tfrac{1}{2} + \nu + \mu)\, \Gamma(\tfrac{1}{2} - \nu + \mu)\, P^{-\mu}_{\nu-\frac{1}{2}}(\cos x) \times$$

$$\times \left\{ \int_0^\pi \frac{f(y)}{\sin y} P^{-\mu}_{\nu-\frac{1}{2}}(-\cos y)\, dy \right\} d\mu, \tag{3.1.63}$$

where $0 < \theta < \pi$, Re $\nu \ge 0$ and $[\text{Re }\nu - \frac{1}{2}]$ is the integral part of $(\text{Re }\nu - \frac{1}{2})$. When $0 \le \text{Re }\nu < \frac{1}{2}$, the empty sum in equation (3.1.63) is assumed to be zero and the

82

representation contains only the last integral term. If Re $\nu < 0$, then ν must be replaced by $-\nu$ and $[\text{Re } \nu - \frac{1}{2}]$ by $[|\text{Re } \nu| - \frac{1}{2}]$ in relation (3.1.63).

Proof:

By applying the ideas used in Chapter 2, the above Theorem 3.1.2 can be established by direct examination of properties of the associated Legendre functions. To prove the expansion theorem, we consider the double integral

$$J(x,T) = \frac{1}{2\pi i} \int_{-iT}^{iT} \mu \, \Gamma\left(\frac{1}{2} + \nu + \mu\right) \Gamma\left(\frac{1}{2} - \nu + \mu\right) P_{\nu-\frac{1}{2}}^{-\mu}(\cos x) \times$$

$$\times \left\{ \int_0^\pi f(y) \, \frac{P_{\nu-\frac{1}{2}}^{-\mu}(-\cos y)}{\sin y} \, dy \right\} d\mu, \qquad (3.1.64)$$

and assume that Re $\nu \geq 0$. The investigation of $J(x,T)$ is similar where Re $\nu < 0$. It follows from the estimates (1.2.25) and (1.2.26) that

$$\int_0^\pi \left| \frac{f(y)}{\sin y} \, P_{\nu-\frac{1}{2}}^{-\mu}(-\cos y) \right| dy \leq O(1) \int_0^a \frac{|f(y)|}{\sin y} \left| \ln \sin \frac{y}{2} \right| dy + O(1) \int_a^\pi \frac{|f(y)|}{\sin y} \, dy.$$

So the condition (ii) of the theorem implies that the inner integral in the equation (3.1.64) converges absolutely and uniformly with respect to μ in the interval $(-iT, iT)$ for arbitrary complex ν. Hence, the integrand of the exterior integral is a continuous function of μ and the double integral is meaningful for arbitrary $T > 0$. Interchanging the order of integration, which is permissible due to uniform convergence of the inner integral, in equation (3.1.64) we can write

$$J(x,T) = \int_0^\pi \frac{f(y)}{\sin y} \, K(x,y,T) \, dy, \qquad (3.1.65)$$

where

$$K(x,y,T) = \frac{1}{2\pi i} \int_{-iT}^{iT} \mu \, \Gamma\left(\frac{1}{2} + \nu + \mu\right) \Gamma\left(\frac{1}{2} - \nu + \mu\right) P_{\nu-\frac{1}{2}}^{-\mu}(\cos x) \, P_{\nu-\frac{1}{2}}^{-\mu}(-\cos y) \, d\mu.$$
$$(3.1.66)$$

Using formula (1.2.19) in the representation (3.1.66), we find that

$$K(x,y,T) = \frac{1}{2\pi i} \int_{-iT}^{iT} \mu \, \Gamma\left(\frac{1}{2} + \nu + \mu\right) \Gamma\left(\frac{1}{2} - \nu + \mu\right) P_{\nu-\frac{1}{2}}^{-\mu}(-\cos x) \, P_{\nu-\frac{1}{2}}^{-\mu}(\cos y) \, d\mu,$$
$$(3.1.67)$$

and for this reason, it is convenient to employ the representation (3.1.66) for $x < y$ and (3.1.67) for $x > y$. The integrands in the representations (3.1.66) and (3.1.67) are analytic functions of the complex variable μ whose only singularities in the half-plane $\operatorname{Re}\mu \geq 0$ are a finite number of poles for $\mu = \nu - \frac{1}{2} - m, m = 0, 1, \ldots, [\operatorname{Re}\nu - \frac{1}{2}]$. Hence, integration along the segment of the imaginary axis can be replaced by integration along a semicircle Γ_T of radius $T > |\nu - \frac{1}{2}|$ in the half-plane $\operatorname{Re}\mu \geq 0$ and using the residue theorem, we obtain

$$K(x,y,T) = \sum_{m=0}^{[\operatorname{Re}\nu-\frac{1}{2}]} (\frac{1}{2}+m-\nu)\,\frac{(-1)^m\,\Gamma(2\nu-m)}{m!}\,P_{\nu-\frac{1}{2}}^{\frac{1}{2}+m-\nu}(\cos y)P_{\nu-\frac{1}{2}}^{\frac{1}{2}+m-\nu}(-\cos x)$$

$$+\frac{1}{2\pi i}\int_{\Gamma_T}\mu\,\Gamma(\frac{1}{2}+\nu+\mu)\,\Gamma(\frac{1}{2}-\nu+\mu)\,P_{\nu-\frac{1}{2}}^{-\mu}(\cos y)\,P_{\nu-\frac{1}{2}}^{-\mu}(-\cos x)\,d\mu, \quad (x>y), \quad (3.1.68)$$

$$K(x,y,T) = \sum_{m=0}^{[\operatorname{Re}\nu-\frac{1}{2}]} (\frac{1}{2}+m-\nu)\frac{(-1)^m\Gamma(2\nu-m)}{m!}\,P_{\nu-\frac{1}{2}}^{\frac{1}{2}+m-\nu}(\cos x)\,P_{\nu-\frac{1}{2}}^{\frac{1}{2}+m-\nu}(-\cos y)$$

$$+\frac{1}{2\pi i}\int_{\Gamma_T}\mu\,\Gamma(\frac{1}{2}+\nu+\mu)\,\Gamma(\frac{1}{2}-\nu+\mu)\,P_{\nu-\frac{1}{2}}^{-\mu}(\cos x)\,P_{\nu-\frac{1}{2}}^{-\mu}(-\cos y)\,d\mu, \quad (x<y). \quad (3.1.69)$$

We now consider the behaviour of $J(x,T)$ as $T \to \infty$. Using the relation (cf. Erdélyi et al. 1953, p. 145)

$$P_{\nu-\frac{1}{2}}^{\frac{1}{2}+m-\nu}(-\cos x) = (-1)^m\,P_{\nu-\frac{1}{2}}^{\frac{1}{2}+m-\nu}(\cos x),$$

together with equations (3.1.68) and (3.1.69), $J(x,T)$ can be written as

$$J(x,T) = \int_0^x \frac{f(y)}{\sin y}\,K_1(x,y,T)\,dy + \int_x^\pi \frac{f(y)}{\sin y}\,K_2(x,y,T)\,dy$$

$$+\sum_{m=0}^{[\operatorname{Re}\nu-\frac{1}{2}]}(\frac{1}{2}+m-\nu)\,\frac{\Gamma(2\nu-m)}{m!}\,P_{\nu-\frac{1}{2}}^{\frac{1}{2}+m-\nu}(\cos x)\int_0^\pi \frac{f(y)}{\sin y}\,P_{\nu-\frac{1}{2}}^{\frac{1}{2}+m-\nu}(\cos y)\,dy$$

$$= J_1(x,T) + J_2(x,T) + \sum_{m=0}^{[\operatorname{Re}\nu-\frac{1}{2}]}(\frac{1}{2}+m-\nu)\,\frac{\Gamma(2\nu-m)}{m!}P_{\nu-\frac{1}{2}}^{\frac{1}{2}+m-\nu}(\cos x)\times$$

$$\times\int_0^\pi \frac{f(y)}{\sin y}\,P_{\nu-\frac{1}{2}}^{\frac{1}{2}+m-\nu}(\cos y)\,dy, \quad (3.1.70)$$

where $K_1(x,y,T)$ and $K_2(x,y,T)$ denote the second terms in the right sides of the representations (3.1.68) and (3.1.69) respectively.

Putting $\mu = T\exp(i\phi)$, $-\frac{\pi}{2} \le \phi \le \frac{\pi}{2}$, and using the asymptotic expansion formula (1.2.45) as $|\mu| \to \infty$, asymptotic expansions for the gamma functions and the inequality (2.1.13), it is found that

$$K_1(x,y,T) = \frac{1}{\pi}\frac{\sin(T\ \ln(\tan\frac{y}{2}/\tan\frac{x}{2}))}{\ln(\tan\frac{y}{2}/\tan\frac{x}{2})}$$

$$+O(1)\frac{1 - exp[-T\ \ln(\tan\frac{x}{2}/\tan\frac{y}{2})]}{T\ \ln(\tan\frac{x}{2}/\tan\frac{y}{2})}, \quad y < x,$$

$$K_2(x,y,T) = \frac{1}{\pi}\frac{\sin(T\ \ln(\tan\frac{x}{2}/\tan\frac{y}{2}))}{\ln(\tan\frac{x}{2}/\tan\frac{y}{2})}$$

$$+O(1)\frac{1 - exp[-T\ \ln(\tan\frac{y}{2}/\tan\frac{x}{2})]}{T\ \ln(\tan\frac{y}{2}/\tan\frac{x}{2})}, \quad x < y. \tag{3.1.71}$$

Substituting $\xi = \ln\tan\frac{x}{2}$ and $\eta = \ln\tan\frac{y}{2}$ and using the relation for $K_1(x,y,T)$ in (3.1.71), $J_1(x,T)$ becomes

$$J_1(x,T) = \frac{1}{\pi}\int_{-\infty}^{\xi} f(\varsigma)\frac{\sin T(\eta - \xi)}{\eta - \xi}\,d\eta$$

$$+O(1)\int_{-\infty}^{\xi}|f(\varsigma)|\frac{1 - exp[-(\xi - \eta)T]}{T(\xi - \eta)}\,d\eta, \tag{3.1.72}$$

where $\varsigma = 2\tan^{-1}(e^\eta)$.

By virtue of the conditions imposed on $f(x)$, it follows that

$$\lim_{T\to\infty}\frac{1}{\pi}\int_{-\infty}^{\xi} f(\varsigma)\frac{\sin T(\eta - \xi)}{\eta - \xi}\,d\eta = \frac{1}{2}f(x - 0). \tag{3.1.73}$$

Further, dividing the range of integration into $(-\infty, \xi - \delta)$ and $(\xi - \delta, \xi)$ and choosing a sufficiently small positive δ and a sufficiently large T, we obtain

$$\int_{-\infty}^{\xi}|f(\varsigma)|\frac{1 - exp[-(\xi - \eta)T]}{T(\xi - \eta)}\,d\eta \le \frac{1}{\delta T}\int_{-\infty}^{\xi - \delta}|f(\varsigma)|\,d\eta + \int_{\xi - \delta}^{\xi}|f(\varsigma)|\,d\eta$$

$$= O(T^{-1}) + o(1) = o(1) \quad \text{as } T \to \infty.$$

Hence,

$$\lim_{T\to\infty} J_1(x,T) = \frac{1}{2}f(x - 0). \tag{3.1.74}$$

85

Similarly, it can be proved that

$$\lim_{T \to \infty} J_2(x, T) = \frac{1}{2} f(x + 0).$$ (3.1.75)

Thus, using the above two relations (3.1.74) and (3.1.75), we find that

$$\lim_{T \to \infty} J(x, T) = \frac{1}{2} [f(x + 0) + f(x - 0)]$$

$$+ \sum_{m=0}^{[\text{Re } \nu - \frac{1}{2}]} (\frac{1}{2} + m - \nu) \frac{\Gamma(2\nu - m)}{m!} P_{\nu - \frac{1}{2}}^{\frac{1}{2} + m - \nu}(\cos x) \int_0^\pi \frac{f(y)}{\sin y} P_{\nu - \frac{1}{2}}^{\frac{1}{2} + m - \nu}(\cos y) \, dy.$$ (3.1.76)

This completes the proof of Theorem 3.1.2.

Belichenko (1987b) used the integral expansion formula (3.1.63) in the solution of a problem of diffraction of electromagnetic waves by wedge with anisotropically conducting faces.

Example 3.1.9

Belichenko (1987a) used the expansion formula (3.1.63) to obtain integral expansions of $\frac{\partial}{\partial \psi} P_{\nu - \frac{1}{2}}(-\cos \beta)$ and $P_{\nu - \frac{1}{2}}(-\cos \beta)$ where $\cos \beta = \cos x \, \cos y + \sin x \, \sin y \, \cos(\phi - \psi)$, $x \in (0, \pi)$, $y \in (0, \pi)$ and $\phi, \psi \in (0, 2\pi)$ being fixed parameters with $|\text{Re } \nu| < \frac{1}{2}$. For this purpose, we first evaluate the integral

$$\int_0^\pi \frac{\partial}{\partial \psi} P_{\nu - \frac{1}{2}}(-\cos \beta) \frac{P_{\nu - \frac{1}{2}}^{-\mu}(-\cos x)}{\sin x} \, dx,$$ (3.1.77)

where μ is purely imaginary. The following addition formula for $P_{\nu - \frac{1}{2}}(-\cos \beta)$ is needed here (cf. Gradshteyn and Ryzhik, 1980, p. 1014):

$$\frac{\partial}{\partial \psi} P_{\nu - \frac{1}{2}}(-\cos \beta) = 2 \sum_{m=1}^\infty (-1)^m \, m \, \frac{\Gamma(\frac{1}{2} + m + \nu)}{\Gamma(\frac{1}{2} - m + \nu)} P_{\nu - \frac{1}{2}}^{-m}(-\cos x_>) \times$$

$$\times P_{\nu - \frac{1}{2}}^{-m}(\cos x_<) \sin m(\phi - \psi),$$ (3.1.78)

where $x_<$ and $x_>$ denote the larger and smaller value of x and y. Multiplying equation (3.1.78) by $P_{\nu - \frac{1}{2}}^{-\mu}(-\cos x)/\sin x$ and then integrating between 0 to π, we obtain the integrals

$$I_m = P_{\nu-\frac{1}{2}}^{-m}(-\cos y) \int_0^y \frac{P_{\nu-\frac{1}{2}}^{-m}(\cos x)\, P_{\nu-\frac{1}{2}}^{-\mu}(-\cos x)}{\sin x}\, dx$$

$$+ P_{\nu-\frac{1}{2}}^{-m}(\cos y) \int_y^\pi \frac{P_{\nu-\frac{1}{2}}^{-m}(-\cos x)\, P_{\nu-\frac{1}{2}}^{-\mu}(-\cos x)}{\sin x}\, dx, \quad m = 1, 2, \ldots$$

$$= \frac{2\cos\pi\nu}{\pi}\, \frac{(-1)^m\, P_{\nu-\frac{1}{2}}^{-\mu}(-\cos y)}{m^2 - \mu^2}\, \frac{\Gamma(\frac{1}{2} - m + \nu)}{\Gamma(\frac{1}{2} + m + \nu)}, \quad |\mathrm{Re}\,\mu| < 1. \qquad (3.1.79)$$

Hence, it is found that

$$\int_0^\pi \frac{\partial}{\partial\psi} P_{\nu-\frac{1}{2}}(-\cos\beta)\, \frac{P_{\nu-\frac{1}{2}}^{-\mu}(-\cos x)}{\sin x}\, dx$$

$$= \pm 2\, P_{\nu-\frac{1}{2}}^{-\mu}(-\cos y)\, \cos\pi\nu\, \frac{\sin[\mu\{\pi \mp (\phi - \psi)\}]}{\sin\mu\pi}, \qquad (3.1.80)$$

where the upper or lower sign is to be taken according as $0 < \phi - \psi < 2\pi$ or $0 < \psi - \phi < 2\pi$ and the relation (3.1.19) has been utilized. Employing the relation (3.1.80) in the expansion formula (3.1.63), we obtain

$$\frac{\partial}{\partial\psi} P_{\nu-\frac{1}{2}}(-\cos\beta) = \mp i\, \cos\pi\nu \int_{-i\infty}^{i\infty} \mu\, \frac{\Gamma(\frac{1}{2} + \nu + \mu)}{\Gamma(\frac{1}{2} + \nu - \mu)}\, \frac{P_{\nu-\frac{1}{2}}^{-\mu}(-\cos y)\, P_{\nu-\frac{1}{2}}^{-\mu}(\cos x)}{\sin\pi\mu\, \cos\pi(\nu - \mu)} \times$$

$$\times \sin[\mu\{\pi \mp (\phi - \psi)\}]d\mu, \quad |\mathrm{Re}\,\nu| < \frac{1}{2}. \qquad (3.1.81)$$

It follows from the above relation that

$$P_{\nu-\frac{1}{2}}(-\cos\beta) = i\, \cos\pi\nu \int_{-i\infty}^{i\infty} \frac{\Gamma(\frac{1}{2} + \nu + \mu)}{\Gamma(\frac{1}{2} + \nu - \mu)}\, \frac{P_{\nu-\frac{1}{2}}^{-\mu}(-\cos y)\, P_{\nu-\frac{1}{2}}^{-\mu}(\cos x)}{\sin\pi\mu\, \cos\pi(\nu - \mu)} \times$$

$$\times \cos[\mu(\pi - |\phi - \psi|)]\, d\mu, \quad |\mathrm{Re}\,\nu| < \frac{1}{2}. \qquad (3.1.82)$$

The integral here is in the sense of the Cauchy Principal Value because the conditions of Theorem 3.1.2 are not satisfied by the function $P_{\nu-\frac{1}{2}}(-\cos\beta)$. It can be easily shown that the relations (3.1.81) and (3.1.82) hold when the contour of integration is closed by a large semicircle in the half-plane $\mathrm{Re}\,\mu \geq 0$ and the integral is evaluated by the residue theorem. It may be noted that x and y can be interchanged in the formulae (3.1.81) and (3.1.82).

3.2 Integral expansions in (x_1, x_2) where $-1 \leq x_1 < x_2 \leq 1$

In Section 3.1, we have seen how an integral expansion for different intervals can be formally developed. Here, we establish two integral expansion formulae for functions defined in a finite interval (x_1, x_2) where $-1 \leq x_1 < x_2 \leq 1$.

3.2.1 For the first integral expansion, we use the notation $\cos\theta$ instead of x and the interval is $(\cos\alpha, 1)$, $0 < \alpha < \pi$. This has been obtained by Mandal and Guha Roy (1991) from the solution of an appropriately designed boundary value problem, in a manner somewhat similar to the method used in Subsection 3.1.2. The boundary value problem involves the determination of a function $u(r, \theta, \phi)$ satisfying

$$(\nabla^2 + k^2)\, u(r, \theta, \phi) = 0, \ 0 < r < \infty, 0 < \theta < \alpha, 0 < \phi < \beta,$$

$$u = 0, \text{ on } \theta = 0, \alpha,$$

$$u = h(r, \theta), \text{ on } \phi = 0,$$

and

$$u = 0, \text{on } \phi = \beta, \tag{3.2.1}$$

where r, θ, ϕ denote spherical polar coordinates and $h(r, \theta)$ is a prescribed function of r and θ.

To solve this boundary value problem, we make use of the Kontorovich–Lebedev transform in the form (cf. Jones, 1964, p. 608)

$$v(\nu, \theta, \phi) = \int_0^\infty r^{-\frac{1}{2}}\, u(r, \theta, \phi)\, H_\nu^{(2)}(kr)\, dr, \ |\text{Re } \nu| < 1,$$

with the inversion formula

$$u(r, \theta, \phi) = -\frac{1}{2} \int_{-i\infty}^{i\infty} \nu\, r^{-\frac{1}{2}}\, J_\nu(kr)\, v(\nu, \theta, \phi)\, d\nu.$$

Then v satisfies the boundary value problem described by

$$\frac{1}{\sin\theta} \frac{\partial}{\partial\theta}\left(\sin\theta \frac{\partial v}{\partial\theta}\right) + \frac{1}{\sin^2\theta} \frac{\partial^2 v}{\partial\phi^2} + \left(\nu^2 - \frac{1}{4}\right)v = 0, \ 0 < \theta < \alpha, \ 0 < \theta < \beta,$$

$$v = 0 \qquad \text{on} \quad \theta = 0, \alpha,$$

$$v = f(\theta) \qquad \text{on} \quad \phi = 0, \qquad\qquad (3.2.2)$$

$$v = 0 \qquad \text{on} \quad \phi = \beta,$$

where $f(\theta)$ is the Kontorovich-Lebedev transform of $h(r, \theta)$.

Let

$$v_n(\theta) = \int_0^\beta v(\nu, \theta, \phi) \sin \lambda\phi \, d\phi, \quad \lambda = \frac{n\pi}{\beta}, \qquad\qquad (3.2.3)$$

so that

$$v = \frac{2}{\beta} \sum_{n=1}^\infty v_n(\theta) \sin \lambda\phi. \qquad\qquad (3.2.4)$$

Then we find that $v_n(\theta)$ satisfies the ordinary differential equation

$$\frac{d}{d\theta}\left(\sin\theta \frac{dv_n}{d\theta}\right) - \frac{\lambda^2}{\sin\theta}v_n + \left(\nu^2 - \frac{1}{4}\right)\sin\theta\, v_n = -\frac{\lambda\, f(\theta)}{\sin\theta}, \quad 0 < \theta < \alpha. \qquad (3.2.5)$$

The Green's function $G(\theta, \theta')$ corresponding to the differential equation (3.2.5) with the boundary conditions $G = 0$ on $\theta = 0$ and α, is given by

$$G(\theta, \theta') = -\frac{\Gamma(\frac{1}{2} - \nu + \lambda)\Gamma(\frac{1}{2} + \nu + \lambda)}{2}\left[\frac{P_{\nu-\frac{1}{2}}^{-\lambda}(-\cos\alpha)}{P_{\nu-\frac{1}{2}}^{-\lambda}(\cos\alpha)} \, P_{\nu-\frac{1}{2}}^{-\lambda}(\cos\theta) \times\right.$$

$$\left. \times\, P_{\nu-\frac{1}{2}}^{-\lambda}(\cos\theta') - P_{\nu-\frac{1}{2}}^{-\lambda}(-\cos\theta)\, P_{\nu-\frac{1}{2}}^{-\lambda}(\cos\theta')\right], \quad \theta < \theta',$$

$$\equiv M_1(\lambda) - M_2(\lambda), \quad \text{say}, \qquad\qquad (3.2.6)$$

where the meaning of the notation M_1 and M_2 is obvious. For $\theta > \theta'$, θ and θ' are to be interchanged.

Now we prove that $G(\theta, \theta')$ has the integral representation given by

$$G(\theta, \theta') = \frac{1}{2\pi i} \int_L \mu\, \frac{[M_1(\mu) - M_2(\mu)]}{\mu^2 - \lambda^2} \, d\mu, \qquad\qquad (3.2.7)$$

for $\theta < \theta'$ and $\theta > \theta'$, where L is the path Re $\mu = c$, parallel to the imaginary axis in the complex μ-plane and $0 < c < \lambda$. To prove this we first consider the integral

$$I_1 \equiv \frac{1}{2\pi i} \int_{c-i\infty}^{c+i\infty} \mu\, \frac{M_1(\mu)}{\mu^2 - \lambda^2} \, d\mu. \qquad\qquad (3.2.8)$$

89

For large $|\mu|$, the integrand in I_1 vanishes for both $\theta < \theta'$ and $\theta > \theta'$, so that by closing the contour on the right side by a large semicircular arc and noting the simple pole of the integrand, we obtain

$$I_1 = M_1(\lambda), \quad \text{for } \theta < \theta' \text{ and } \theta > \theta'. \tag{3.2.9}$$

Again, for the integral

$$I_2 \equiv \frac{1}{2\pi i} \int_{c-i\infty}^{c+i\infty} \mu \, \frac{M_2(\mu)}{\mu^2 - \lambda^2} \, d\mu,$$

we note that the integrand vanishes for $\theta > \theta'$ as Re $\mu \to \infty$, so that closing the contour on the right side, we find as in I_1 that

$$I_2 = M_2(\lambda), \quad \text{for } \theta > \theta'. \tag{3.2.10}$$

To show that (3.2.10) also holds for $\theta < \theta'$, we write for convenience $I_2 \equiv A(\theta, \theta')$. Since the integrand in I_2, i.e., $A(\theta, \theta')$, has no singularity between the imaginary axis and the line Re $\mu = c$, we can take $c = 0$. Then we have

$$A(\theta, \theta') - A(\theta', \theta) = \int_{-i\infty}^{i\infty} a(\mu) \, d\mu,$$

where $a(\mu)$ is the difference of the integrands in $A(\theta, \theta')$ and $A(\theta', \theta)$. It is obvious that $a(\mu)$ is an odd function of μ so that

$$\int_{-i\infty}^{i\infty} a(\mu) \, d\mu = 0,$$

and hence $A(\theta, \theta') = A(\theta', \theta)$. Thus (3.2.10) also holds for $\theta < \theta'$. Therefore, the integral representation (3.2.7) holds for both $\theta > \theta'$ and $\theta < \theta'$.

Now the solution of the differential equation (3.2.5) is

$$v_n(\theta) = -\lambda \int_0^a \frac{f(\theta')}{\sin \theta'} G(\theta, \theta') \, d\theta'. \tag{3.2.11}$$

Using the representation (3.2.7), equation (3.2.11) becomes

$$v_n(\theta) = -\frac{\lambda}{2\pi i} \int_{c-i\infty}^{c+i\infty} \mu \, \Gamma\left(\frac{1}{2} - \nu + \mu\right) \Gamma\left(\frac{1}{2} + \nu + \mu\right) P_{\nu-\frac{1}{2}}^{-\mu}(\cos\theta) \frac{F(\mu)}{\mu^2 - \lambda^2} \, d\mu,$$

where

$$F(\mu) = \int_0^\alpha \left[\frac{P_{\nu-\frac{1}{2}}^{-\mu}(-\cos\alpha)}{P_{\nu-\frac{1}{2}}^{-\mu}(\cos\alpha)} P_{\nu-\frac{1}{2}}^{-\mu}(\cos\theta) - P_{\nu-\frac{1}{2}}^{-\mu}(-\cos\theta) \right] \frac{f(\theta)}{\sin\theta} \, d\theta. \qquad (3.2.12)$$

Using relation (3.2.3), we find that

$$v(\theta,\phi) = \frac{1}{2\pi i} \int_{c-i\infty}^{c+i\infty} \mu\Gamma\left(\frac{1}{2}-\nu+\mu\right)\Gamma\left(\frac{1}{2}+\nu+\mu\right) \frac{\sin(\beta-\phi)\mu}{\sin\beta\mu} P_{\nu-\frac{1}{2}}^{-\mu}(\cos\theta) \, F(\mu) d\mu. \qquad (3.2.13)$$

Again, utilizing the boundary condition that $v(\theta,0) = f(\theta)$, from the relation (3.2.13), we obtain

$$f(\theta) = \frac{1}{2\pi i} \int_{c-i\infty}^{c+i\infty} \mu\,\Gamma\left(\frac{1}{2}-\nu+\mu\right)\Gamma\left(\frac{1}{2}+\nu+\mu\right) P_{\nu-\frac{1}{2}}^{-\mu}(\cos\theta) \, F(\mu) \, d\mu. \qquad (3.2.14)$$

Thus formulae (3.2.12) and (3.2.14) constitute the required expansion formula for the function $f(\theta)$ defined on $(0,\alpha)$.

If we now make $\alpha = \pi$, formula (3.2.14) reduces to (3.1.50), (3.1.22) and (3.1.32) (after writing $c = 0$) as discussed in the previous Section 3.1.

Example 3.2.1

Taking $f(\theta) = (\sin\theta)^\nu$ where $\nu = -\frac{1}{2} + i\tau$ and using the above expansion formula (3.2.14), we find the following integral expansion formula for $(\sin\theta)^\nu$ as

$$(\sin\theta)^\nu = \frac{2^\nu\,\Gamma(1+\nu)}{2\pi i} \int_{c-i\infty}^{c+i\infty} \frac{\mu}{\mu+\nu} \Gamma\left(\frac{1}{2}-\nu+\mu\right)\Gamma\left(\frac{1}{2}+\nu+\mu\right) P_\nu^{-\mu}(\cos\theta)\times$$

$$\times \left[\frac{P_\nu^{-\mu}(\cos\alpha)}{P_\nu^{-\mu}(-\cos\alpha)} P_\nu^{-\nu}(\cos\alpha) P_{\nu-1}^{-\mu}(\cos\alpha) - P_\nu^{-\nu}(\cos\alpha) P_{\nu-1}^{-\mu}(-\cos\alpha) \right] d\mu.$$

In the derivation of this integral expansion use of a result given in Erdélyi et al. 1953, p. 169 has been utilized.

3.2.2   A second integral expansion has been established by Mandal and Mandal (1994). They proved rigorously this integral expansion for a class of functions defined on a finite interval (x_1, x_2) where $-1 \le x_1 < x_2 \le 1$. It may be noted that this expansion can be developed formally from the δ-function representation given by the relation (3.1.8). The main result is given in the form of the following theorem.

Theorem 3.2.1

Let $f(x)$ be a given function defined on the interval (x_1, x_2) where $-1 \le x_1 < x_2 \le 1$ and satisfying the following conditions:

(i) $f(x)$ is piecewise continuous and has bounded variation in (x_1, x_2),

(ii) $f(x)(1 - x^2)^{-1} \ln(1 - x^2)^{-1} \in L(x_1, x_2)$, $-1 \le x_1 < x_2 \le 1$.

Then at the points of continuity of $f(x)$, we have

$$f(x) = \sum_k \sigma_k \Gamma\left(\frac{1}{2} + i\tau - i\sigma_k\right) \Gamma\left(\frac{1}{2} - i\tau - i\sigma_k\right) \frac{M(x, x_2; i\sigma_k)}{(\partial/\partial\sigma_k)M(x_2, x_1; i\sigma_k)} F(\sigma_k)$$

$$+ \frac{1}{2\pi i} \int_{-\infty}^{\infty} \sigma \Gamma\left(\frac{1}{2} + i\tau - i\sigma\right) \Gamma\left(\frac{1}{2} - i\tau - i\sigma\right) \frac{M(x, x_2; i\sigma)}{M(x_2, x_1; i\sigma)} F(\sigma)\, d\sigma, \qquad (3.2.15)$$

with

$$F(\sigma) = \int_{x_1}^{x_2} \frac{f(x)}{1 - x^2} M(x, x_1; i\sigma)\, d\sigma, \quad -1 \le x_1 < x_2 \le 1, \qquad (3.2.16)$$

where $M(x, y; i\sigma) = P^{i\sigma}_{-\frac{1}{2}+i\tau}(x) P^{i\sigma}_{-\frac{1}{2}+i\tau}(-y) - P^{i\sigma}_{-\frac{1}{2}+i\tau}(-x) P^{i\sigma}_{-\frac{1}{2}+i\tau}(y)$; the σ_k's, σ, τ are real and the σ_k's satisfy $M(x_2, x_1; i\sigma_k) = 0$.

The equation (3.2.16) may be regarded as an integral transform of the function $f(x)$ defined on (x_1, x_2) and (3.2.15) is its inverse. The formulae (3.2.15) and (3.2.16) together give the integral expansion of the function $f(x)$.

Proof:

The theorem is proved here in a manner similar to Theorem 2.1.2. Using the estimate (1.2.24), it follows from equation (3.2.16) that

$$\int_{x_1}^{x_2} \left| \frac{f(x)}{1 - x^2} \left[P^{i\sigma}_{-\frac{1}{2}+i\tau}(x) P^{i\sigma}_{-\frac{1}{2}+i\tau}(-x_1) - P^{i\sigma}_{-\frac{1}{2}+i\tau}(-x) P^{i\sigma}_{-\frac{1}{2}+i\tau}(x_1) \right] \right| dx$$

$$\le \sqrt{\frac{\sinh \pi\sigma}{\pi\sigma}} \int_{x_1}^{x_2} \frac{|f(x)|}{1 - x^2} \left\{ P_{-\frac{1}{2}+i\tau}(x) P_{-\frac{1}{2}+i\tau}(-x_1) + P_{-\frac{1}{2}+i\tau}(-x) P_{-\frac{1}{2}+i\tau}(x_1) \right\} dx.$$

This shows that the conditions imposed on $f(x)$ imply that the integral $F(\sigma)$ is absolutely and uniformly convergent for $\sigma \in [-T, T]$ where T is a positive large number. Hence, $F(\sigma)$ is continuous on $[-T, T]$ and the repeated integral

$$J(x,T) = \frac{1}{2\pi i} \int_{-T}^{T} \sigma \, \Gamma\left(\frac{1}{2} + i\tau - i\sigma\right) \Gamma(\frac{1}{2} - i\tau - i\sigma) \frac{M(x,x_2;i\sigma)}{M(x_2,x_1;i\sigma)} \times$$

$$\times \left\{ \int_{x_1}^{x_2} \frac{f(y)}{1-y^2} M(y,x_1;i\sigma) \, dy \right\} d\sigma \qquad (3.2.17)$$

is meaningful. Also, uniform convergence allows us to change the order of integration and write $J(x,T)$ as

$$J(x,T) = \int_{x_1}^{x_2} \frac{f(y)}{1-y^2} K(x,y,T) \, dy, \qquad (3.2.18)$$

where

$$K(x,y,T) = \frac{1}{2\pi i} \int_{-T}^{T} \sigma \, \Gamma\left(\frac{1}{2} + i\tau - i\sigma\right) \Gamma\left(\frac{1}{2} - i\tau - i\sigma\right) \frac{M(x,x_2;i\sigma)M(y,x_1;i\sigma)}{M(x_2,x_1;i\sigma)} \, d\sigma.$$
$$(3.2.19)$$

It is simple to show that the kernel $K(x,y,T)$ is symmetric in the variables x and y. Thus

$$K(x,y,T) = K(y,x,T). \qquad (3.2.20)$$

Now we consider the behaviour of the kernel $K(x,y,T)$ as $T \to \infty$. Putting $\mu = -i\sigma$, we write (3.2.19) as

$$K(x,y,T) = \frac{1}{2\pi i} \int_{-iT}^{iT} \mu \, \Gamma\left(\frac{1}{2} + i\tau + \mu\right) \Gamma\left(\frac{1}{2} - i\tau + \mu\right) \frac{M(x,x_2;-\mu)M(y,x_1;-\mu)}{M(x_2,x_1;-\mu)} d\mu.$$
$$(3.2.21)$$

The integrand of $K(x,y,T)$ is an analytic function of the complex variable μ and it has a singularity in the half-plane $\mathrm{Re}\,\mu \geq 0$ except for the simple poles at $\mu = -i\sigma_k$ $(k = 1,2,\ldots)$, where

$$M(x_2,x_1;i\sigma_k) = 0, \quad \sigma_k > 0. \qquad (3.2.22)$$

We can, therefore, replace the integration along the imaginary axis by that on the large semicircle Γ_T of radius T in the half-plane $\mathrm{Re}\,\mu \geq 0$ and apply the residue theorem. Then

$$K(x,y,T) = K_1(x,y,T) - \sum_k \sigma_k \, \Gamma\left(\frac{1}{2} + i\tau - i\sigma_k\right) \Gamma\left(\frac{1}{2} - i\tau - i\sigma_k\right) \times$$

$$\times \frac{M(x,x_2;i\sigma_k)\, M(y,x_1;i\sigma_k)}{(\partial/\partial\sigma_k)M(x_2,x_1;i\sigma_k)}, \tag{3.2.23}$$

where

$$K_1(x,y,T) = \frac{1}{2\pi i}\int_{\Gamma_T} \mu\Gamma\left(\frac{1}{2} + i\tau + \mu\right)\Gamma\left(\frac{1}{2} - i\tau + \mu\right)\frac{M(x,x_2;-\mu)\, M(y,x_1;-\mu)}{M(x_2,x_1;-\mu)}\, d\mu. \tag{3.2.24}$$

We now suppose that $y \le x$. By virtue of the estimates (1.2.44) and asymptotic properties of the gamma function for large $|\mu|$, we conclude that

$$\mu\,\Gamma\left(\frac{1}{2} + i\tau + \mu\right)\Gamma\left(\frac{1}{2} - i\tau + \mu\right)\frac{M(x,x_2;-\mu)\, M(y,x_1;-\mu)}{M(x_2,x_1;-\mu)}$$

$$= \frac{\left[\left(\dfrac{1+x}{1-x}\dfrac{1-x_2}{1+x_2}\right)^{-\frac{\mu}{2}} - \left(\dfrac{1-x}{1+x}\dfrac{1+x_2}{1-x_2}\right)^{-\frac{\mu}{2}}\right]}{\left[\left(\dfrac{1+x_2}{1-x_2}\dfrac{1-x_1}{1+x_1}\right)^{-\frac{\mu}{2}} - \left(\dfrac{1-x_2}{1+x_2}\dfrac{1+x_1}{1-x_1}\right)^{-\frac{\mu}{2}}\right]} \times$$

$$\times \left[\left(\frac{1+y}{1-y}\frac{1-x_1}{1+x_1}\right)^{-\frac{\mu}{2}} - \left(\frac{1-y}{1+y}\frac{1+x_1}{1-x_1}\right)^{-\frac{\mu}{2}}\right]\left[1 + O(|\mu|^{-1})\right]. \tag{3.2.25}$$

Now we introduce the new variables

$$\xi = \frac{1}{2}\ln\frac{1+x}{1-x}, \quad \eta = \frac{1}{2}\ln\frac{1+y}{1-y}, \quad \alpha = \frac{1}{2}\ln\frac{1+x_1}{1-x_1} \quad \text{and} \quad \beta = \frac{1}{2}\ln\frac{1+x_2}{1-x_2}.$$

Then, for $y \le x$, we find

$$K_1(x,y,T) = \frac{1}{2\pi i}\int_{\Gamma_T}\left[\exp\{-\mu(\xi-\eta)\} + \exp\{-\mu(2\beta - 2\alpha - \xi + \eta)\} -\right.$$

$$\exp\{-\mu(\xi+\eta-2\alpha)\} - \exp\{-\mu(2\beta - \xi - \eta)\}\right]d\mu + O(1)\int_0^{\pi/2}\left[\exp\{-\mu(\xi-\eta)\cos\phi\}\right.$$

$$+\exp\{-\mu(2\beta - 2\alpha - \xi + \eta)\cos\phi\} - \exp\{-\mu(\xi+\eta-2\alpha)\cos\phi\}$$

$$-\exp\{-\mu(2\beta - \xi - \eta)\cos\phi\}\right]d\phi, \quad \alpha < \eta \le \xi < \beta. \tag{3.2.26}$$

Using the inequality (2.1.13), (3.2.26) becomes

$$K_1(x,y,T) = \frac{1}{\pi}\left[\frac{\sin T(\xi-\eta)}{\xi-\eta} + \frac{\sin T(2\beta-2\alpha-\xi+\eta)}{2\beta-2\alpha-\xi+\eta}\right.$$

$$\left.-\frac{\sin T(\xi+\eta-2\alpha)}{\xi+\eta-2\alpha} - \frac{\sin T(2\beta-\xi-\eta)}{2\beta-\xi-\eta}\right]$$

$$+O(1)\left[\frac{1-\exp\{-T(\xi-\eta)\}}{T(\xi-\eta)} + \frac{1-\exp\{-T(2\beta-2\alpha-\xi+\eta)\}}{T(2\beta-2\alpha-\xi+\eta)}\right.$$

$$\left.-\frac{1-\exp\{-T(\xi+\eta-2\alpha)\}}{T(\xi+\eta-2\alpha)} - \frac{1-\exp\{-T(2\beta-\xi-\eta)\}}{T(2\beta-\xi-\eta)}\right], \quad \alpha<\eta\le\xi<\beta,$$

$$(3.2.27)$$

where the factor $O(1)$ is independent of y.

Again for $y \ge x$, we use the symmetry property (3.2.20) and the representation (3.2.24) with the variables x,y replaced by y,x.

Now we write the representation (3.2.18) as

$$\begin{aligned}
J(x,T) &= \int_{z_1}^{x} \frac{f(y)}{1-y^2} K_1(x,y,T)\,dy + \int_{x}^{z_2} \frac{f(y)}{1-y^2} K_1(x,y,T)\,dy \\
&\quad - \sum_k \sigma_k \, \Gamma\left(\frac{1}{2}+i\tau-i\sigma_k\right) \Gamma\left(\frac{1}{2}-i\tau-i\sigma_k\right) \frac{M(x,x_2;i\sigma_k)}{(\partial/\partial\sigma_k)M(x_2,x_1;i\sigma_k)} \times \\
&\quad \times \int_{z_1}^{z_2} \frac{f(y)}{1-y^2} M(y,x_1;i\sigma_k)\,dy \\
&= J_1(x,T)+J_2(x,T) - \sum_k \sigma_k \, \Gamma\left(\frac{1}{2}+i\tau-i\sigma_k\right) \Gamma\left(\frac{1}{2}-i\tau-i\sigma_k\right) \times \\
&\quad \times \frac{M(x,x_2;i\sigma_k)}{(\partial/\partial\sigma_k)M(x_2;x_1;i\sigma_k)} \int_{z_1}^{z_2} \frac{f(y)}{1-y^2} M(y,x_1;i\sigma_k)\,dy. \quad (3.2.28)
\end{aligned}$$

Using relation (3.2.27) in J_1, we obtain

$$\begin{aligned}
J_1(x,T) &= \frac{1}{\pi}\left[\int_{\alpha}^{\xi} f(\tanh\eta)\,\frac{\sin T(\xi-\eta)}{(\xi-\eta)}\,d\eta\right. \\
&\quad + \int_{\alpha}^{\xi} f(\tanh\eta)\,\frac{\sin T(2\beta-2\alpha-\xi+\eta)}{(2\beta-2\alpha-\xi+\eta)}\,d\eta \\
&\quad - \int_{\alpha}^{\xi} f(\tanh\eta)\,\frac{\sin T(\xi+\eta-2\alpha)}{(\xi+\eta-2\alpha)}\,d\eta \\
&\quad \left.- \int_{\alpha}^{\xi} f(\tanh\eta)\,\frac{\sin T(2\beta-\xi-\eta)}{(2\beta-\xi-\eta)}\,d\eta\right]
\end{aligned}$$

95

$$+ O(1) \left[\int_\alpha^\xi |f(\tanh \eta)| \frac{1 - \exp\{-T(\xi - \eta)\}}{T(\xi - \eta)} \, d\eta \right.$$

$$+ \int_\alpha^\xi |f(\tanh \eta)| \frac{1 - \exp\{-T(2\beta - 2\alpha - \xi + \eta)\}}{T(2\beta - 2\alpha - \xi + \eta)} \, d\eta$$

$$- \int_\alpha^\xi |f(\tanh \eta)| \frac{1 - \exp\{-T(\xi + \eta - 2\alpha)\}}{T(\xi + \eta - 2\alpha)} \, d\eta$$

$$\left. - \int_\alpha^\xi |f(\tanh \eta)| \frac{1 - \exp\{-T(2\beta - \xi - \eta)\}}{T(2\beta - \xi - \eta)} \, d\eta \right]. \tag{3.2.29}$$

The conditions satisfied by $f(x)$ imply that $f(\tanh \eta) \in L(-1, 1)$, hence, as $T \to \infty$, we find

$$\frac{1}{\pi} \int_\alpha^\xi f(\tanh \eta) \frac{\sin T(\xi - \eta)}{(\xi - \eta)} \, d\eta = \frac{1}{2} f(\tanh \xi - 0) + o(1)$$

$$= \frac{1}{2} f(x - 0) + o(1),$$

$$\frac{1}{\pi} \int_\alpha^\xi f(\tanh \eta) \frac{\sin T(2\beta - 2\alpha - \xi + \eta)}{(2\beta - 2\alpha - \xi + \eta)} \, d\eta = o(1),$$

$$\frac{1}{\pi} \int_\alpha^\xi f(\tanh \eta) \frac{\sin T(\xi + \eta - 2\alpha)}{(\xi + \eta - 2\alpha)} \, d\eta = o(1),$$

and

$$\frac{1}{\pi} \int_\alpha^\xi f(\tanh \eta) \frac{\sin T(2\beta - \xi - \eta)}{2\beta - \xi - \eta} \, d\eta = o(1). \tag{3.2.30}$$

Moreover, if the interval of integration is divided into the subintervals $(\xi - \delta, \xi)$ and $(\alpha, \xi - \delta)$ and if a sufficiently small positive δ (implying a sufficiently large T) is chosen, then we have

$$\int_\alpha^\xi |f(\tanh \eta)| \frac{1 - \exp\{-T(\xi - \eta)\}}{T(\xi - \eta)} \, d\eta \leq \frac{1}{\delta T} \int_\alpha^{\xi - \delta} |f(\tanh \eta)| \, d\eta$$

$$+ \int_{\xi - \delta}^\xi |f(\tanh \eta)| \, d\eta$$

$$= O(T^{-1}) + o(1) = o(1) \quad \text{as } T \to \infty,$$

$$\int_\alpha^\xi |f(\tanh \eta)| \frac{1 - \exp\{-T(2\beta - 2\alpha - \xi + \eta)\}}{T(2\beta - 2\alpha - \xi + \eta)} \, d\eta \leq \frac{1}{\xi T} \int_\alpha^\xi |f(\tanh \eta)| \, d\eta$$

$$= O(T^{-1}) = o(1) \quad \text{as } T \to \infty,$$

96

$$\int_\alpha^\xi |f(\tanh\eta)| \frac{1-\exp\{-T(\xi+\eta-2\alpha)\}}{T(\xi+\eta-2\alpha)} \, d\eta \le \frac{1}{\xi T} \int_\alpha^\xi |f(\tanh\eta)| \, d\eta$$

$$= O(T^{-1}) = o(1) \quad \text{as } T \to \infty,$$

and

$$\int_\alpha^\xi |f(\tanh\eta)| \frac{1-\exp\{-T(2\beta-\xi-\eta)\}}{T(2\beta-\xi-\eta)} \, d\eta \le \frac{1}{\xi T} \int_\alpha^\xi |f(\tanh\eta)| \, d\eta$$

$$= O(T^{-1}) = o(1) \quad \text{as } T \to \infty. \tag{3.2.31}$$

Thus relations (3.2.29) – (3.2.31) lead to

$$\lim_{T\to\infty} J_1(\tanh\xi, T) = \frac{1}{2} f(\tanh\xi - 0) = \frac{1}{2} f(x-0). \tag{3.2.32}$$

Similarly,

$$\lim_{T\to\infty} J_2(\tanh\xi, T) = \frac{1}{2} f(\tanh\xi + 0) = \frac{1}{2} f(x+0). \tag{3.2.33}$$

Hence,

$$\lim_{T\to\infty} J(x, T) = \frac{1}{2}[f(x+0) + f(x-0)] - \sum_k \sigma_k \, \Gamma\left(\frac{1}{2} + i\tau - i\sigma_k\right) \Gamma\left(\frac{1}{2} - i\tau - i\sigma_k\right) \times$$

$$\times \frac{M(x, x_2; i\sigma_k)}{(\partial/\partial\sigma_k)M(x_2, x_1; i\sigma_k)} \, F(\sigma_k). \tag{3.2.34}$$

This proves Theorem 3.2.1. It may be observed that when $x_1 = -1$ and $x_2 = 1$, the expansion formula (3.2.15) reduces to the formula (3.1.32).

It follows from Theorem 3.2.1 that, at the points of continuity of $f(x)$, we have

$$f(x) = \sum_k \sigma_k \, \Gamma\left(\frac{1}{2} + i\tau - i\sigma_k\right) \Gamma\left(\frac{1}{2} - i\tau - i\sigma_k\right) \frac{R(x, x_2; i\sigma_k)}{(\partial^2/\partial x_2 \partial\sigma_k)R(x_2, x_1; i\sigma_k)} \, F(\sigma_k)$$

$$+ \frac{1}{2\pi i} \int_{-\infty}^{\infty} \sigma \, \Gamma\left(\frac{1}{2} + i\tau - i\sigma\right) \Gamma\left(\frac{1}{2} - i\tau - i\sigma\right) \frac{R(x, x_2; i\sigma)}{(\partial/\partial x_2)R(x_2, x_1; i\sigma)} \, F(\sigma) \, d\sigma, \tag{3.2.35}$$

with

$$F(\sigma) = \int_{x_1}^{x_2} \frac{f(x)}{(1-x^2)} R(x, x_1; i\sigma) \, dx, \quad -1 < x_1 < x_2 < 1, \tag{3.2.36}$$

97

where

$$R(x,y,i\sigma) = P^{i\sigma}_{-\frac{1}{2}+i\tau}(x)\frac{\partial}{\partial y}P^{i\sigma}_{-\frac{1}{2}+i\tau}(-y) - P^{i\sigma}_{-\frac{1}{2}+i\tau}(-x)\frac{\partial}{\partial y}P^{i\sigma}_{-\frac{1}{2}+i\tau}(y),$$

and the σ_k's, σ, τ are real numbers. The integrand in (3.2.35) has singularities at $\sigma = \sigma_k$ $(k = 1, 2, \ldots)$ which are simple poles along the positive σ-axis, where

$$\frac{\partial}{\partial x_2}R(x_2, x_1; i\sigma_k) = 0, \quad \sigma_k > 0.$$

To prove the expansion formula (3.2.35), we have to use the following estimates for large $|\mu|$:

$$\frac{d}{dx}P^{-\mu}_{-\frac{1}{2}+i\tau}(x) = -\frac{\mu}{\Gamma(1+\mu)}\frac{1}{(1-x)(1+x)}\left(\frac{1+x}{1-x}\right)^{-\frac{\mu}{2}}[1+O(|\mu|^{-1})],$$

$$\frac{d}{dx}P^{-\mu}_{-\frac{1}{2}+i\tau}(-x) = -\frac{\mu}{\Gamma(1+\mu)}\frac{1}{(1+x)(1-x)}\left(\frac{1-x}{1+x}\right)^{-\frac{\mu}{2}}[1+O(|\mu|^{-1})].$$

The proof of formula (3.2.35) is similar to the proof of the foregoing Theorem 3.2.1. We note here that formula (3.2.35) reduces to formula (3.1.32) when $x_1 = -1$ and $x_2 = 1$.

Below we give integral expansions of two simple functions as illustrations of the above integral expansion theorem.

Example 3.2.2

$$(1-x^2)^{\frac{\nu}{2}} = \sum_k \sigma_k \, \Gamma\left(\frac{1}{2} + i\tau - i\sigma_k\right) \, \Gamma\left(\frac{1}{2} - i\tau - i\sigma_k\right) \frac{M(x, x_2; i\sigma_k)}{(\partial/\partial\sigma_k)M(x_2, x_1; i\sigma_k)} \times$$

$$\times \frac{2^\nu \, \Gamma(1+\nu)}{(\nu^2 + \sigma_k^2)}(\nu + i\sigma_k)\left[P^{-\nu}_\nu(x_1)M_1(x_1, x_1; i\sigma_k) - P^{-\nu}_\nu(x_2)M_1(x_2, x_1; i\sigma_k)\right]$$

$$+\frac{2^\nu \, \Gamma(1+\nu)}{2\pi i}\int_{-\infty}^\infty \frac{\sigma(\nu + i\sigma)}{(\nu^2 + \sigma^2)}\Gamma\left(\frac{1}{2} + i\tau - i\sigma\right)\Gamma\left(\frac{1}{2} - i\tau - i\sigma\right)\frac{M(x, x_2; i\sigma)}{M(x_2, x_1; i\sigma)} \times$$

$$\times \left[P^{-\nu}_\nu(x_1)\, M_1(x_1, x_1; i\sigma) - P^{-\nu}_\nu(x_2)\, M_1(x_2, x_1; i\sigma)\right] d\sigma, -1 < x_1 < x_2 < 1,$$

where

$$M_1(x, y; i\sigma) = P^{i\sigma}_{\nu-1}(x)\, P^{i\sigma}_\nu(-y) - P^{i\sigma}_{\nu-1}(-x)\, P^{i\sigma}_\nu(y), \quad \nu = -\frac{1}{2} + i\tau.$$

Example 3.2.3

$$P_\nu^\mu(x) = \sum_k \sigma_k \, \Gamma\left(\frac{1}{2} + i\tau - i\sigma_k\right) \Gamma\left(\frac{1}{2} - i\tau - i\sigma_k\right) \frac{M(x, x_2; i\sigma_k)}{(\partial/\partial\sigma_k)M(x_2, x_1; i\sigma_k)} \frac{1}{(\mu^2 + \sigma_k^2)} \times$$

$$\times \left[(\nu + \mu)P_{\nu-1}^\mu(x_2)M(x_2, x_1; i\sigma_k) + (\nu + i\sigma_k)\{P_\nu^\mu(x_1)M_1(x_1, x_2; i\sigma_k)\right.$$

$$\left. -P_\nu^\mu(x_2)M_1(x_2, x_2; i\sigma_k)\}\right] + \frac{1}{2\pi i} \int_{-\infty}^{\infty} \frac{\sigma}{(\mu^2 + \sigma^2)} \Gamma\left(\frac{1}{2} + i\tau - i\sigma\right) \Gamma\left(\frac{1}{2} - i\tau - i\sigma\right) \times$$

$$\times \frac{M(x, x_2; i\sigma)}{M(x_2, x_1; i\sigma)} \left[(\nu + \mu)P_{\nu-1}^\mu(x_2)M(x_2, x_1; i\sigma) + (\nu + i\sigma)\{P_\nu^\mu(x_1)M_1(x_1, x_2; i\sigma)\right.$$

$$\left. -P_\nu^\mu(x_2)\, M_1(x_2, x_1; i\sigma)\}\right] d\sigma.$$

It is not difficult to ascertain that in these examples the conditions under which the expansion theorem holds are satisfied.

3.3 Integral expansions in $(1, \infty)$

In this section, we obtain the integral expansion formula for functions defined on $(1, \infty)$ involving associated Legendre functions whose argument lies in the interval $(1, \infty)$.

3.3.1 Mandal (1971b) developed an integral expansion formula of this type from the consideration of a boundary value problem involving the part of a hemisphere in the toroidal coordinate system (α, β, ϕ) defined by

$$x = \frac{c \sinh\alpha \cos\phi}{\cosh\alpha + \cos\beta}, \quad y = \frac{c \sinh\alpha \sin\phi}{\cosh\alpha + \cos\beta}, \quad z = \frac{c \sin\beta}{\cosh\alpha + \cos\beta}.$$

The problem is to find $u = u(\alpha, \beta, \phi)$ satisfying the Laplace equation (in toroidal coordinates)

$$\frac{\partial}{\partial\alpha}\left(\frac{\sinh\alpha}{\cosh\alpha + \cos\beta}\frac{\partial u}{\partial\alpha}\right) + \frac{\partial}{\partial\beta}\left(\frac{\sinh\alpha}{\cosh\alpha + \cos\beta}\frac{\partial u}{\partial\beta}\right) + \frac{\operatorname{cosech}\alpha}{(\cosh\alpha + \cos\beta)}\frac{\partial^2 u}{\partial\phi^2} = 0,$$

$$(3.3.1)$$

in the region $0 \le \alpha < \infty, 0 \le \beta \le \beta_0(< \pi), 0 \le \phi \le \phi_0$ along with the boundary conditions on $\phi = 0, \phi_0$, as given by

$$u(\alpha, \beta, \phi) = \begin{cases} f(\cosh\alpha) (\cosh\alpha + \cos\beta)^{\frac{1}{2}} e^{\pm\tau\beta}, & on \ \phi = 0, \\ 0, & on \ \phi = \phi_0, \end{cases} \quad (3.3.2)$$

99

where r is real and fixed, while on the other boundary surfaces, u may be prescribed in any manner and u is required to be finite everywhere.

Substituting $s = \cosh \alpha$ and $u = (\cosh \alpha + \cos \beta)^{\frac{1}{2}} e^{\pm r\beta} v(s, \phi)$ in equation (3.3.1), we see that v satisfies the partial differential equation

$$\frac{\partial}{\partial s}\left[(s^2 - 1)\frac{\partial v}{\partial s}\right] + \left(\frac{1}{4} + r^2\right)v + \frac{1}{s^2 - 1}\frac{\partial^2 v}{\partial \phi^2} = 0, \quad 1 \leq s < \infty, \tag{3.3.3}$$

along with the boundary conditions

$$\left.\begin{array}{c} v(s, 0) = f(s), \quad 1 \leq s < \infty, \\ v(s, \phi_0) = 0, \quad 1 \leq s < \infty. \end{array}\right\} \tag{3.3.4}$$

If we represent $v(s, \phi)$ as

$$v(s, \phi) = \frac{2}{\phi_0}\sum_{n=1}^{\infty} v_n(s) \sin \lambda\phi, \quad \lambda = \frac{n\pi}{\phi_0}, \tag{3.3.5}$$

then

$$v_n(s) = \int_0^{\phi_0} v(s, \phi) \sin \lambda\phi \, d\phi, \quad 1 \leq s < \infty. \tag{3.3.6}$$

Multiplying both sides of equation (3.3.3) by $\sin \lambda\phi$, then integrating with respect to ϕ and by using the relations (3.3.5) and (3.3.6), we observe that $\psi(s) \equiv v_n(s)$ satisfies the ordinary differential equation

$$\frac{d}{ds}\left[(s^2 - 1)\frac{d\psi}{ds}\right] + \left(\frac{1}{4} + r^2\right)\psi - \frac{\lambda^2}{s^2 - 1}\psi = -\frac{\lambda f(s)}{s^2 - 1}, \quad 1 \leq s < \infty, \tag{3.3.7}$$

along with the boundary conditions

$$v_n(s) = 0, \quad \text{on } s = 1,$$

and

$$v_n(s) \text{ is finite as } s \to \infty. \tag{3.3.8}$$

The solution of the differential equation (3.3.7) is given by

$$\psi(s) = -\lambda \int_1^{\infty} \frac{f(s')}{s'^2 - 1} G(s, s') \, ds', \tag{3.3.9}$$

100

where $G(s, s')$ is the Green's function satisfying the homogeneous associated Legendre equation

$$\frac{d}{ds}\left[(s^2 - 1)\frac{d\psi}{ds}\right] + \left(\frac{1}{4} + \tau^2 - \frac{\lambda^2}{s^2 - 1}\right)\psi = 0, \quad 1 \le s < \infty, \tag{3.3.10}$$

with the boundary conditions (3.3.8).

Using two independent solutions of the associated Legendre equation (3.3.10) as $P^{-\lambda}_{-\frac{1}{2}+i\tau}(s)$ and $Q^{\lambda}_{-\frac{1}{2}+i\tau}(s)$, the Green's function $G(s, s')$ is obtained as

$$G(s, s') = -e^{-i\pi\lambda}\, P^{-\lambda}_{-\frac{1}{2}+i\tau}(s_<)\, Q^{\lambda}_{-\frac{1}{2}+i\tau}(s_>)$$

where $s_<, s_>$ denote respectively the smaller and greater of s, s'. Now we consider the integral

$$\frac{1}{\pi i}\int_{\Gamma_T} e^{-i\pi\mu}\, P^{-\mu}_{-\frac{1}{2}+i\tau}(s_<)\, Q^{\mu}_{-\frac{1}{2}+i\tau}(s_>)\, \frac{\mu}{\mu^2 - \lambda^2}\, d\mu, \tag{3.3.11}$$

where Γ_T is a large semicircle of radius $T\ (> 0)$ in the anticlockwise sense in the half-plane $\operatorname{Re}\mu \ge 0$ with its centre at $\mu = c\ (c < \frac{\pi}{\phi_0})$. From the hypergeometric properties of the associated Legendre functions $P^{-\mu}_{-\frac{1}{2}+i\tau}(\cosh\alpha)$ and $Q^{\mu}_{-\frac{1}{2}+i\tau}(\cosh\alpha)$, we find that the integrand of expression (3.3.11) has no singularity except at the simple pole at $\mu = \lambda$ inside the contour Γ_T. Using the asymptotic relations (1.2.46) and (1.2.47), as $|\mu| \to \infty$ in $\operatorname{Re}\mu > 0$, the integrand in expression (3.3.11) is approximately equal to the expression

$$\frac{1}{2i\pi\mu^2}\left(\frac{\tanh\frac{\alpha_<}{2}}{\tanh\frac{\alpha_>}{2}}\right)^{\mu}\left[e^{\pm\pi i(1+\nu)}\left(\frac{\mu}{1+\nu-\mu}\right)^{1+\nu} + \frac{e^{(1+\nu-\mu)\operatorname{sech}^2\frac{\alpha_>}{2}}}{(1+\nu-\mu)^{1+\nu}}\right]\coth\frac{\alpha_>}{2}.$$

Now $\tanh\frac{\alpha}{2}$ is a monotonic increasing function of α in $(0, \infty)$ so that $\left(\frac{\tanh\frac{\alpha_<}{2}}{\tanh\frac{\alpha_>}{2}}\right) < 1$. Hence, the integrand tends to zero as $|\mu| \to \infty$. So the integrand along the arc of the semicircle of Γ_T vanishes. Thus

$$G(s, s') = \frac{1}{\pi i}\int_{c-i\infty}^{c+i\infty} e^{-i\pi\mu}\, P^{-\mu}_{-\frac{1}{2}+i\tau}(s_<)\, Q^{\mu}_{-\frac{1}{2}+i\tau}(s_>)\, \frac{\mu}{\mu^2 - \lambda^2}\, d\mu. \tag{3.3.12}$$

To prove the uniqueness of the representation (3.3.12), we have to show

$$\int_{c-i\infty}^{c+i\infty} e^{-i\pi\mu}\left[P^{-\mu}_{-\frac{1}{2}+i\tau}(s')\, Q^{\mu}_{-\frac{1}{2}+i\tau}(s) - P^{-\mu}_{-\frac{1}{2}+i\tau}(s)\, Q^{\mu}_{-\frac{1}{2}+i\tau}(s')\right]\frac{\mu}{\mu^2 - \lambda^2}\, d\mu = 0.$$

Let $I(\mu)$ denote the expression within the square bracket of the integrand. Now we consider the integral

$$\int_L e^{-i\pi\mu} I(\mu) \frac{\mu}{\mu^2 - \lambda^2} d\mu$$

where L is the contour bounded by the lines parallel to the imaginary axis at a distance c from it and lying on the left and on the right of it, and being closed at infinity by lines parallel to the real axis in the complex μ-plane. Since $\lambda = \frac{n\pi}{\phi_0}$ and $c < \frac{\pi}{\phi_0}$, there is no singularity inside the contour L. Hence, the integral vanishes. There will be no contribution to the integral from the lines parallel to the real axis as $I(-\mu) = \exp(-2i\pi\mu) I(\mu)$. Thus

$$\int_{c-i\infty}^{c+i\infty} e^{-i\pi\mu} I(\mu) \frac{\mu}{\mu^2 - \lambda^2} d\mu + \int_{-c-i\infty}^{-c+i\infty} e^{-i\pi\mu} I(\mu) \frac{\mu}{\mu^2 - \lambda^2} d\mu = 0.$$

Changing μ to $-\mu$ in the second integral, we obtain

$$2 \int_{c-i\infty}^{c+i\infty} e^{-i\pi\mu} I(\mu) \frac{\mu}{\mu^2 - \lambda^2} d\mu = 0.$$

So the representation (3.3.12) of the Green's function $G(s, s')$ is unique.

Substituting the value of G from (3.3.12) in equation (3.3.9), we get

$$\psi(s) = -\frac{\lambda}{\pi i} \int_1^\infty \frac{f(s')}{s'^2 - 1} \left\{ \int_{c-i\infty}^{c+i\infty} e^{-i\pi\mu} P_{-\frac{1}{2}+ir}^{-\mu}(s) Q_{-\frac{1}{2}+ir}^{\mu}(s') \frac{\mu}{\mu^2 - \lambda^2} d\mu \right\} ds'.$$

$$(3.3.13)$$

Assuming the uniform convergence of the above integral, we can change the order of integration and then we find that

$$\psi(s) = -\frac{\lambda}{\pi i} \int_{c-i\infty}^{c+i\infty} e^{-i\pi\mu} P_{-\frac{1}{2}+ir}^{-\mu}(s) \frac{\mu}{\mu^2 - \lambda^2} \left\{ \int_1^\infty \frac{f(s')}{s'^2 - 1} Q_{-\frac{1}{2}+ir}^{\mu}(s') ds' \right\} d\mu.$$

$$(3.3.14)$$

Using relation (3.1.19), from equations (3.3.5) and (3.3.14) it is found that

$$v(s, \phi) = \frac{1}{\pi i} \int_{c-i\infty}^{c+i\infty} e^{-i\pi\mu} P_{-\frac{1}{2}+ir}^{-\mu}(s) \frac{\mu \, \sin\mu(\phi_0 - \phi)}{\sin\mu\phi_0} \times$$

$$\times \left\{ \int_1^\infty \frac{f(s')}{s'^2 - 1} Q_{-\frac{1}{2}+ir}^{\mu}(s') ds' \right\} d\mu. \qquad (3.3.15)$$

Applying the boundary condition (3.3.4), equation (3.3.15) becomes

$$f(s) = \frac{1}{i\pi} \int_{c-i\infty}^{c+i\infty} \mu \, e^{-i\pi\mu} \, P_{-\frac{1}{2}+i\tau}^{-\mu}(s) \, F(\mu) \, d\mu, \qquad (3.3.16)$$

where

$$F(\mu) = \int_1^\infty \frac{f(s)}{s^2-1} \, Q_{-\frac{1}{2}+i\tau}^{\mu}(s) \, ds. \qquad (3.3.17)$$

Formula (3.3.17) gives the integral transform while formula (3.3.16) gives the corresponding inversion and formulae (3.3.16) and (3.3.17) together give the integral expansion formula.

3.3.2   Mandal and Mandal (1993) proved rigorously the above expansion formula (3.3.16) for $c = 0$. The proof of the expansion formula is based on the ideas used in earlier sections. The main result is presented in the form of the following theorem.

Theorem 3.3.1

Let $f(\alpha)$ be a given function defined on $(0, \infty)$ and satisfying the following conditions:

(i) $f(x)$ is piecewise continuous and is of bounded variation in $(0, \infty)$,

(ii) $\dfrac{|f(\alpha)|}{\sinh \alpha} \, Q_{-\frac{1}{2}}^{\sigma}(\cosh \alpha) \in L(0, \infty), \, |\sigma| < \frac{1}{2}$.

Then, at the points of continuity of $f(\alpha)$

$$f(\alpha) = \frac{1}{\pi i} \int_{-i\infty}^{i\infty} \mu \, e^{-i\pi\mu} \, P_{-\frac{1}{2}+i\tau}^{-\mu}(\cosh \alpha) \, F(\mu) \, d\mu, \qquad (3.3.18)$$

where

$$F(\mu) = \int_0^\infty \frac{f(\alpha)}{\sinh \alpha} \, Q_{-\frac{1}{2}+i\tau}^{\mu}(\cosh \alpha) \, d\alpha, \qquad (3.3.19)$$

$\mu = \sigma + ip$ and τ is a real parameter. Equation (3.3.19) may be regarded as an integral transform of the function $f(\alpha)$ defined on $(0, \infty)$ and (3.3.18) is its inverse. Equations (3.3.18) and (3.3.19) together give the integral expansion of the function.

Proof:

Using the inequality (1.2.23), it follows that

$$\int_0^\infty \left| \frac{f(\alpha)}{\sinh \alpha} \, Q_{-\frac{1}{2}+i\tau}^{\mu}(\cosh \alpha) \right| d\alpha \le \frac{\Gamma(\frac{1}{2}-\sigma)}{|\Gamma(\frac{1}{2}-\mu)|} \int_0^\infty \frac{|f(\alpha)|}{\sinh \alpha} \, Q_{-\frac{1}{2}}^{\sigma}(\cosh \alpha) \, d\alpha,$$

103

and this proves that, under the conditions imposed on $f(\alpha)$, the integral $F(\mu)$ is majorized by a convergent integral. So the integral in (3.3.19) is absolutely and uniformly convergent. Hence, $F(\mu)$ represents a continuous function of μ and the double integral

$$J(\alpha,T) = \frac{1}{\pi i} \int_{-iT}^{iT} \mu \, e^{-i\pi\mu} \, P_{-\frac{1}{2}+i\tau}^{-\mu}(\cosh\alpha) \left\{ \int_0^\infty \frac{f(\alpha')}{\sinh\alpha'} \, Q_{-\frac{1}{2}+i\tau}^{\mu}(\cosh\alpha') \, d\alpha' \right\} d\mu$$

has a meaning. Due to uniform convergence, changing the order of integration we write $J(\alpha,T)$ as

$$J(\alpha,T) = \int_0^\infty \frac{f(\alpha')}{\sinh\alpha'} \, K(\alpha,\alpha',T) \, d\alpha', \qquad (3.3.20)$$

where

$$K(\alpha,\alpha',T) = \frac{1}{\pi i} \int_{-iT}^{iT} \mu \, e^{-i\pi\mu} \, P_{-\frac{1}{2}+i\tau}^{-\mu}(\cosh\alpha) \, Q_{-\frac{1}{2}+i\tau}^{\mu}(\cosh\alpha') \, d\mu. \qquad (3.3.21)$$

Utilizing the properties of the associated Legendre functions, it can be shown that the kernel $K(\alpha,\alpha',T)$ is symmetric in the variables α and α'. Thus

$$K(\alpha,\alpha',T) = K(\alpha',\alpha,T). \qquad (3.3.22)$$

Now we consider the behaviour of the kernel $K(\alpha,\alpha',T)$ as $T \to \infty$. The integrand of $K(\alpha,\alpha',T)$ is a function of the complex variable μ and regular in the half-plane $\text{Re}\,\mu \geq 0$. We can, therefore, replace the integration by that on the semicircle Γ_T of radius T in the half-plane $\text{Re}\,\mu \geq 0$. Then

$$K(\alpha,\alpha',T) = \frac{1}{\pi i} \int_{\Gamma_T} \mu \, e^{-i\pi\mu} \, P_{-\frac{1}{2}+i\tau}^{-\mu}(\cosh\alpha) \, Q_{-\frac{1}{2}+i\tau}^{\mu}(\cosh\alpha') \, d\mu. \qquad (3.3.23)$$

Now, we fix α and suppose that $\alpha \leq \alpha'$. Using the asymptotic properties of the associated Legendre functions (1.2.46) and (1.2.47) and the gamma function as $|\mu| \to \infty$, we obtain

$$\mu \, e^{-i\pi\mu} \, P_\nu^{-\mu}(\cosh\alpha) \, Q_\nu^{\mu}(\cosh\alpha') = \frac{1}{2} \left(\frac{\tanh\frac{\alpha}{2}}{\tanh\frac{\alpha'}{2}} \right)^\mu \times$$

$$\times \left[1 + \exp\left\{ \pm i\pi(1+\nu) + (1+\nu-\mu) \, \text{sech}^2\frac{\alpha'}{2} \right\} \right] \left[1 + O(|\mu|^{-1}) \right]. \qquad (3.3.24)$$

104

Putting $\xi = \ln \tanh \frac{\alpha}{2}$ and $\eta = \ln \tanh \frac{\alpha'}{2}$, we obtain from (3.3.23) and (3.3.24) for $\alpha \leq \alpha'$ (i.e. $\xi \leq \eta$) that

$$K(\alpha, \alpha', T) = \frac{1}{2\pi i} \int_{\Gamma_T} \left[\exp\{-\mu(\eta - \xi)\} + \exp\{\pm i\pi(1 + \nu) + (1 + \nu - \mu)\times \right.$$

$$\left. \times (1 - e^{2\eta}) - \mu(\eta - \xi)\}\right] d\mu + O(1) \left[\int_0^{\pi/2} \exp\{-T(\eta - \xi)\cos\phi\} d\phi \right.$$

$$\left. + e^{\pi T} \int_0^{\pi/2} \exp\{-T(\eta - \xi + 1 - e^{2\eta})\cos\phi\} d\phi \right]. \tag{3.3.25}$$

Using the inequality (2.1.13), equation (3.3.25) reduces to the form

$$K(\alpha, \alpha', T) = \frac{1}{\pi} \left[\frac{\sin T(\eta - \xi)}{\eta - \xi} + \exp\{(\pm i\pi + 1 - e^{2\eta})(1 + \nu)\} \times \right.$$

$$\left. \times \frac{\sin T(\eta - \xi + 1 - e^{2\eta})}{\eta - \xi + 1 - e^{2\eta}}\right] + O(1) \left[\frac{1 - \exp\{-T(\eta - \xi)\}}{T(\eta - \xi)} + \right.$$

$$\left. + e^{\pi T} \frac{1 - \exp\{-T(\eta - \xi + 1 - e^{2\eta})\}}{T(\eta - \xi + 1 - e^{2\eta})}\right]. \tag{3.3.26}$$

For $\alpha \geq \alpha'$, use of symmetry property (3.3.22) and the representation (3.3.23), we find that

$$K(\alpha', \alpha, T) = \frac{1}{\pi} \left[\frac{\sin T(\xi - \eta)}{\xi - \eta} + \exp\{(\pm i\pi + 1 - e^{2\xi})(1 + \nu)\} \times \right.$$

$$\left. \times \frac{\sin T(\xi - \eta + 1 - e^{2\xi})}{\xi - \eta + 1 - e^{2\xi}}\right] + O(1) \left[\frac{1 - \exp\{-T(\xi - \eta)\}}{T(\xi - \eta)} + \right.$$

$$\left. + e^{\pi T} \frac{1 - \exp\{-T(\xi - \eta + 1 - e^{2\xi})\}}{T(\xi - \eta + 1 - e^{2\xi})}\right]. \tag{3.3.27}$$

Now we divide $J(\alpha, T)$ into two parts as, say,

$$J(\alpha, T) = \int_0^\alpha \frac{f(\alpha')}{\sinh \alpha'} K(\alpha, \alpha', T) \, d\alpha' + \int_\alpha^\infty \frac{f(\alpha')}{\sinh \alpha'} K(\alpha, \alpha', T) \, d\alpha'$$

$$= J_1(\alpha, T) + J_2(\alpha, T). \tag{3.3.28}$$

Using the relation (3.3.27) in J_1, we obtain

$$J_1(\alpha, T) = \frac{1}{\pi} \int_{-\infty}^{\xi} f(2\tanh^{-1} e^\eta) \, \frac{\sin T(\xi - \eta)}{(\xi - \eta)} \, d\eta$$

$$+ \frac{1}{\pi} \int_{-\infty}^{\xi} f(2\tanh^{-1} e^\eta) \exp\left\{(\pm i\pi + 1 - e^{2\xi})(1 + \nu)\right\} \frac{\sin T(\xi - \eta + 1 - e^{2\xi})}{(\xi - \eta + 1 - e^{2\xi})} \, d\eta$$

$$+ O(1) \int_{-\infty}^{\xi} |f(2\tanh^{-1} e^\eta)| \, \frac{1 - \exp\{-T(\xi - \eta)\}}{T(\xi - \eta)} \, d\eta$$

$$+ O(1) \int_{-\infty}^{\xi} |f(2\tanh^{-1} e^\eta)|e^{\pi\tau} \, \frac{1 - \exp\{-T(\xi - \eta + 1 - e^{2\xi})\}}{T(\xi - \eta + 1 - e^{2\xi})} \, d\eta. \qquad (3.3.29)$$

Since $f(2\tanh^{-1} e^\eta) \in L(-\infty, \infty)$, it follows that, as $T \to \infty$, the first integral in the right side of (3.3.29) tends to $\frac{1}{2}(2\tanh^{-1} e^\xi - 0)$ and the second integral tends to zero. It is easily shown that the third and fourth integrals also tend to zero as $T \to \infty$. Hence,

$$\lim_{T \to \infty} J_1(\alpha, T) = \frac{1}{2} f(\alpha - 0), \qquad (3.2.30)$$

and employing (3.3.26), we similarly obtain

$$\lim_{T \to \infty} J_2(\alpha, T) = \frac{1}{2} f(\alpha + 0). \qquad (3.3.31)$$

Hence, we conclude that

$$\lim_{T \to \infty} J(\alpha, T) = \frac{1}{2} [f(\alpha + 0) + f(\alpha - 0)]. \qquad (3.3.32)$$

At the points of continuity of $f(\alpha)$, the expansion formula (3.3.18) is obtained.

Using a result in Erdélyi et al. (1953), p. 172, examples of the integral expansion formula (3.3.18) for some simple functions are now obtained.

Example 3.3.1

$$(\sinh \alpha)^\lambda = \frac{2^{-\lambda}}{\pi i} \int_{-i\infty}^{i\infty} \mu \, 2^\mu \, \frac{\Gamma(\frac{1}{2} + \frac{1}{2}\nu + \frac{1}{2}\mu) \, \Gamma(1 + \frac{1}{2}\nu - \frac{1}{2}\lambda)}{\Gamma(1 + \frac{1}{2}\nu - \frac{1}{2}\mu) \, \Gamma(\frac{1}{2} + \frac{1}{2}\nu + \frac{1}{2}\lambda)} \times$$

$$\times \Gamma\left(\frac{1}{2}\lambda + \frac{1}{2}\mu\right) \Gamma\left(\frac{1}{2}\lambda - \frac{1}{2}\mu\right) P_\nu^{-\mu}(\cosh \alpha) \, d\mu, \qquad (3.3.33)$$

where $0 < \operatorname{Re} \lambda < \frac{3}{2}$.

Example 3.3.2

$$P_\lambda^{-\lambda}(\cosh \alpha) = \frac{2^{-2\lambda}}{i\pi\Gamma(1+\lambda)} \int_{-i\infty}^{i\infty} \mu \, 2^\mu \, \frac{\Gamma(\frac{1}{2} + \frac{1}{2}\nu + \frac{1}{2}\mu) \, \Gamma(1 + \frac{1}{2}\nu - \frac{1}{2}\lambda)}{\Gamma(1 + \frac{1}{2}\nu - \frac{1}{2}\mu) \, \Gamma(\frac{1}{2} + \frac{1}{2}\nu + \frac{1}{2}\lambda)} \times$$

$$\times \Gamma\left(\frac{1}{2}\lambda + \frac{1}{2}\mu\right) \Gamma\left(\frac{1}{2}\lambda - \frac{1}{2}\mu\right) P_\nu^{-\mu}(\cosh \alpha) \, d\mu, \quad 0 < \mathrm{Re}\,\lambda < \frac{3}{2}. \qquad (3.3.34)$$

Example 3.3.3

Another interesting example is

$$1 = 2i \int_{-i\infty}^{i\infty} 2^\mu \, \mathrm{cosec}\left(\frac{\pi\mu}{2}\right) \frac{\Gamma(\frac{1}{2} + \frac{1}{2}\nu + \frac{1}{2}\mu) \, \Gamma(1 + \frac{1}{2}\nu)}{\Gamma(1 + \frac{1}{2}\nu - \frac{1}{2}\mu) \, \Gamma(\frac{1}{2} + \frac{1}{2}\nu)} P_\nu^{-\mu}(\cosh \alpha) \, d\mu, \qquad (3.3.35)$$

which cannot be obtained directly from the expansion formula (3.3.18) because the interior integral diverges. Formally, Example 3.3.3 can be obtained from the previous Example 3.3.1 by letting $\lambda \to 0$.

3.4 Integral expansion in $(1, a)$ $(a > 1)$

We consider here the integral expansion of functions $f(x)$ defined on a finite interval $(1, a)$ $(a > 1)$ involving associated Legendre functions in which the argument lies in $(1, a)$. If we take $x = \cosh \alpha$, then α lies in the interval $(0, \beta)$ where $a = \cosh \beta$ and for the sake of notational convenience, we write $f(\alpha)$ for $f(\cosh \alpha)$. The corresponding integral expansion theorem is the following.

Theorem 3.4.1

Let $f(\alpha)$ be a given function defined on $(0, \beta)$, $\beta > 0$, and satisfying the following conditions:

(i) $f(x)$ is piecewise continuous and possesses bounded variation in $(0, \beta)$,

(ii) $\dfrac{|f(\alpha)|}{\sinh \alpha} \, P_{-\frac{1}{2}}^{-\sigma}(\cosh \alpha) \in L(0, \beta)$, $\sigma \geq 0, \beta > 0$.

Then at the points of continuity of $f(\alpha)$,

$$f(\alpha) = \frac{1}{2\pi i} \int_{-i\infty}^{i\infty} \mu \, \Gamma\left(\frac{1}{2} + i\tau + \mu\right) \Gamma\left(\frac{1}{2} - i\tau + \mu\right) P_{-\frac{1}{2}+i\tau}^{-\mu}(\cosh \alpha) \, F(\mu) \, d\mu, \qquad (3.4.1)$$

where

$$F(\mu) = \int_0^\beta \left[P_{-\frac{1}{2}+i\tau}^{-\mu}(-\cosh\alpha) - \frac{P_{-\frac{1}{2}+i\tau}^{-\mu}(-\cosh\beta)}{P_{-\frac{1}{2}+i\tau}^{-\mu}(\cosh\beta)} \, P_{-\frac{1}{2}+i\tau}^{-\mu}(\cosh\alpha) \right] \frac{f(\alpha)}{\sinh\alpha} \, d\alpha,$$

(3.4.2)

and $\mu = \sigma + ip$, $\beta > 0$ and τ is a given real parameter. Equation (3.4.2) may be regarded as an integral transform of the function $f(\alpha)$ defined on $(0, \beta)$ and equation (3.4.1) is its inverse.

This theorem has been established rigorously by Mandal and Mandal (1992). Using the inequalities (1.2.22), (1.2.23), asymptotic expansion formula (1.2.46) and following the method used in the proof of Theorem 3.3.1, this theorem can be established.

We now give examples of integral representation (3.4.1) of some particular functions. Using a relation of Erdélyi et al. (1953), p. 169, we can establish the following integral representations of some simple functions.

Example 3.4.1

$$(\sinh\alpha)^\nu = \frac{2^\nu \, \Gamma(1+\nu)}{2\pi i} \, P_\nu^{-\nu}(\cosh\beta) \int_{-i\infty}^{i\infty} \mu \, \frac{\Gamma(1+\nu+\mu) \, \Gamma(\mu-\nu)}{(\nu+\mu)} \, P_\nu^{-\mu}(\cosh\alpha)$$

$$\times \left[\frac{P_\nu^{-\mu}(-\cosh\beta)}{P_\nu^{-\mu}(\cosh\beta)} \, P_{\nu-1}^{-\mu}(\cosh\beta) - P_{\nu-1}^{-\mu}(-\cosh\beta) \right] d\mu,$$

where $\nu = -\frac{1}{2} + i\tau$.

Example 3.4.2

$$P_\nu^{-\mu'}(\cosh\alpha) = \frac{1}{2\pi i} \int_{-i\infty}^{i\infty} \frac{\mu}{(\mu'^2 - \mu^2)} \, \Gamma(1+\nu+\mu) \, \Gamma(\mu-\nu) \, P_\nu^{-\mu}(\cosh\alpha) \times$$

$$\times \left[(\nu+\mu') \, P_\nu^{-\mu}(\cosh\beta) \, P_{\nu-1}^{-\mu'}(\cosh\beta) - (\nu-\mu) \, P_{\nu-1}^{-\mu}(\cosh\beta) \times \right.$$

$$\times P_\nu^{\mu'}(\cosh\beta) - \frac{P_\nu^{-\mu}(-\cosh\beta)}{P_\nu^{-\mu}(\cosh\beta)} \left\{ (\nu+\mu') P_\nu^{-\mu}(\cosh\beta) \times \right.$$

$$\left. \left. \times P_{\nu-1}^{\mu'}(\cosh\beta) - (\nu-\mu) \, P_{\nu-1}^{-\mu}(\cosh\beta) \, P_\nu^{\mu'}(\cosh\beta) \right\} \right] d\mu,$$

where $\nu = -\frac{1}{2} + i\tau$.

Example 3.4.3

$$\frac{\cosh \tau\alpha}{\sqrt{\sinh \alpha}} = \frac{1}{i}\sqrt{\frac{2}{\pi}}\int_{-i\infty}^{i\infty}\mu\,\frac{\Gamma(\frac{1}{2}+\mu+i\tau)\,\Gamma(\frac{1}{2}+\mu-i\tau)}{(1-4\mu^2)}\,P_{-\frac{1}{2}+i\tau}^{-\mu}(\cosh\alpha)\,\times$$

$$\times\left[i\tau P_{-\frac{1}{2}+i\tau}^{-\mu}(-\cosh\beta)\,P_{-\frac{1}{2}+i\tau}^{\frac{1}{2}}(\cosh\beta) - (i\tau - \mu - \frac{1}{2})\times\right.$$

$$\times P_{-\frac{1}{2}+i\tau}^{-\mu}(-\cosh\beta)\,P_{-\frac{1}{2}+i\tau}^{\frac{1}{2}}(\cosh\beta) - \frac{P_{-\frac{1}{2}+i\tau}^{-\mu}(-\cosh\beta)}{P_{-\frac{1}{2}+i\tau}^{-\mu}(\cosh\beta)}\times$$

$$\times\left\{i\tau P_{-\frac{1}{2}+i\tau}^{-\mu}(\cosh\beta)\,P_{-\frac{1}{2}+i\tau}^{\frac{1}{2}}(\cosh\beta) - (i\tau-\mu-\frac{1}{2})\times\right.$$

$$\left.\left.\times\,P_{-\frac{1}{2}+i\tau}^{-\mu}(\cosh\beta)\,P_{-\frac{1}{2}+i\tau}^{\frac{1}{2}}(\cosh\beta)\right\}\right]d\mu.$$

Example 3.4.4

$$\frac{\sinh \tau\alpha}{\sqrt{\sinh \alpha}} = -i\tau\sqrt{\frac{2}{\pi}}\int_{-i\infty}^{i\infty}\mu\,\frac{\Gamma(\frac{1}{2}+\mu+i\tau)\,\Gamma(\frac{1}{2}+\mu-i\tau)}{(1-4\mu^2)}\,P_{-\frac{1}{2}+i\tau}^{-\mu}(\cosh\alpha)\,\times$$

$$\times\left[(i\tau-1)P_{-\frac{1}{2}+i\tau}^{-\mu}(-\cosh\beta)\,P_{-\frac{1}{2}+i\tau}^{-\frac{1}{2}}(\cosh\beta) - (i\tau - \mu - \frac{1}{2})\right.$$

$$\times P_{-\frac{1}{2}+i\tau}^{-\mu}(-\cosh\beta)\,P_{-\frac{1}{2}+i\tau}^{-\mu}(\cosh\beta) - \frac{P_{-\frac{1}{2}+i\tau}^{-\mu}(-\cosh\beta)}{P_{-\frac{1}{2}+i\tau}^{-\mu}(\cosh\beta)}\times$$

$$\times\left\{(i\tau-1)P_{-\frac{1}{2}+i\tau}^{-\mu}(\cosh\beta)\,P_{-\frac{1}{2}+i\tau}^{-\frac{1}{2}}(\cosh\beta) - (i\tau-\mu-\frac{1}{2})\times\right.$$

$$\left.\left.\times\,P_{-\frac{1}{2}+i\tau}^{-\mu}(\cosh\beta)\,P_{-\frac{1}{2}+i\tau}^{-\frac{1}{2}}(\cosh\beta)\right\}\right]d\mu.$$

Example 3.4.5

$$\frac{e^{-i\tau a}}{\sqrt{\sinh \alpha}} = -\left(\frac{2}{\pi}\right)^{\frac{3}{2}}\int_{-i\infty}^{i\infty}\mu\,\frac{\Gamma(\frac{1}{2}+\mu+i\tau)\,\Gamma(\frac{1}{2}+\mu-i\tau)}{(1-4\mu^2)}\,P_{-\frac{1}{2}+i\tau}^{-\mu}(\cosh\alpha)\,\times$$

$$\times\left[i\tau P_{-\frac{1}{2}+i\tau}^{-\mu}(-\cosh\beta)\,Q_{-\frac{1}{2}+i\tau}^{\frac{1}{2}}(\cosh\beta) - (i\tau - \mu - \frac{1}{2})\,\times\right.$$

$$\times P_{-\frac{1}{2}+i\tau}^{-\mu}(-\cosh\beta)\,Q_{-\frac{1}{2}+i\tau}^{\frac{1}{2}}(\cosh\beta) - \frac{P_{-\frac{1}{2}+i\tau}^{-\mu}(-\cosh\beta)}{P_{-\frac{1}{2}+i\tau}^{-\mu}(\cosh\beta)}\times$$

$$\times\left\{i\tau P_{-\frac{1}{2}+i\tau}^{-\mu}(\cosh\beta)\,Q_{-\frac{1}{2}+i\tau}^{\frac{1}{2}}(\cosh\beta) - (i\tau-\mu-\frac{1}{2})\,\times\right.$$

$$\left.\left.\times\,P_{-\frac{1}{2}+i\tau}^{-\mu}(\cosh\beta)\,Q_{-\frac{1}{2}+i\tau}^{\frac{1}{2}}(\cosh\beta)\right\}\right]d\mu.$$

Chapter 4

Integral expansions related to Mehler–Fock type transforms involving generalized associated Legendre functions

A number of integral expansion theorems involving generalized associated Legendre functions have been established rigorously by Braaksma and Meulenbeld (1967). The generalized associated Lengendre functions $P_k^{\mu,\nu}(z)$ and $Q_k^{\mu,\nu}(z)$ involve two superscripts. In Section 2.4, we have presented integral expansions involving these functions where the subscript appears as an integration variable while the superscripts remain fixed. Here we present integral expansions where any one of the superscripts appears as an integration variable while the other superscript and the subscript remain fixed. For proofs of these expansions, the reader is referred to the original paper of Braaksma and Meulenbeld (1967). These expansion formulae are also utilized to obtain integral expansions of some simple functions.

4.1 Integral expansions in $(-1, 1)$

Theorem 4.1.1

Let γ be a real number with

$$\gamma < \min\left\{\operatorname{Re}(2k + 2 - \mu), \operatorname{Re}(-2k - \mu)\right\}, \qquad (4.1.1)$$

and $f(x)$ be a function such that for all $a, -1 < a < 1$

$$\left.\begin{aligned}
f(x)(1 + x)^{-\frac{1}{4} - \frac{1}{2}|\operatorname{Re}\mu|} &\in L(-1, a), \ \operatorname{Re}\mu \neq 0, \\
f(x)(1 + x)^{-\frac{1}{4}}\ln(1 + x) &\in L(-1, a), \ \operatorname{Re}\mu = 0, \\
f(x)(1 - x)^{-1 - \frac{1}{2}\gamma} &\in L(a, 1).
\end{aligned}\right\} \qquad (4.1.2)$$

Further, let $f(x)$ be of bounded variation in the open interval $(-1, 1)$. Then $f(x)$ has the following integral expansions:

$$\frac{1}{2}[f(x+0) + f(x-0)] = -\frac{1}{4\pi i} \int_{\gamma-i\infty}^{\gamma+i\infty} \nu \, \Gamma(k - \frac{\mu+\nu}{2} + 1) \, \Gamma(-k - \frac{\mu+\nu}{2}) \times$$

$$\times P_k^{\mu,\nu}(-x) \left\{ \int_{-1}^{1} \frac{f(y)}{1-y} \, P_k^{\nu,\mu}(y) \, dy \right\} d\nu, \tag{4.1.3}$$

and

$$\frac{1}{2}[f(x+0) + f(x-0)] = -\frac{1}{4\pi i} \int_{\gamma-i\infty}^{\gamma+i\infty} \nu \, \Gamma(k - \frac{\mu+\nu}{2} + 1) \, \Gamma(-k - \frac{\mu+\nu}{2}) \times$$

$$\times P_k^{\nu,\mu}(x) \left\{ \int_{-1}^{1} \frac{f(y)}{1-y} \, P_k^{\mu,\nu}(-y) \, dy \right\} d\nu. \tag{4.1.4}$$

Theorem 4.1.2

Let γ be a real number satisfying the inequality (4.1.1), γ_0 be a complex number with Re $\gamma_0 \leq 0$, and

$$\text{Re } \gamma_0 < \min \{ \text{Re}(2k + 2 - \mu), \text{Re}(-2k - \mu) \}. \tag{4.1.5}$$

Suppose that $f(z)$ is a function continuous on Re $z \leq \max(\gamma, \text{Re } \gamma_0)$, and analytic on Re $z < \max(\gamma, \text{Re } \gamma_0)$. Further, let

$$z^{\frac{1}{2}+\mu} f(z) \in L(\gamma - i\infty, \gamma + i\infty), \tag{4.1.6}$$

and as $|z| \to \infty$

$$f(z) = \begin{cases} 2^{-\frac{1}{2}z} \, z^{-1-|\text{Re } \mu|} \, o(1), & \text{Re } \mu < 1 \ (\mu \neq 0), \\ \dfrac{2^{-\frac{1}{2}z}}{z \, \ln z} \, o(1), & \mu = 0, \\ 2^{-\frac{1}{2}z} \, z^{1-3\mu} \, o(1), & \text{Re } \mu > 1, \\ \dfrac{2^{-\frac{1}{2}z}}{z^2 \, \ln z} \, o(1), & \text{Re } \mu = 1. \end{cases}$$

on Re $z \leq \max \{\gamma, \text{Re } \gamma_0\}$.

If Re $\gamma_0 = \gamma \leq 0$, then $f(z)$ has to satisfy a Hölder condition in a neighbourhood of γ_0. Then $f(\gamma_0)$ has the following integral expansion:

$$f(\gamma_0) = -\frac{1}{4\pi i} \, \Gamma(k - \frac{\mu+\gamma_0}{2}) \, \Gamma(-k - \frac{\mu+\gamma_0}{2}) \int_{-1}^{1} \frac{1}{1-y} \, P_k^{\gamma_0,\mu}(y) \times$$

$$\times \left\{ \int_{\gamma-i\infty}^{\gamma+i\infty} \nu \, f(\nu) \, P_k^{\mu,\nu}(-y) \, d\nu \right\} dy. \tag{4.1.7}$$

111

Theorem 4.1.3

Let μ be a complex number with $\operatorname{Re} \mu < 1$. Let S be the strip $|\operatorname{Re} z| < a$ in the complex z-plane, and $\overline{S}$ be the strip $|\operatorname{Re} z| \leq a$, where a is a positive number such that $\Gamma(k - \frac{\mu+z}{2} + 1)\, \Gamma(-k - \frac{\mu+z}{2})$ has no poles in $\overline{S}$. Let γ be a real number and γ_0 be a complex number in $\overline{S}$.

Let $f(z)$ be a function analytic in S, and continuous in $\overline{S}$ satisfying the conditions:

(i) $\qquad f(z) = 2^z f(-z),$ $\hfill (4.1.8)$

(ii) $\quad z^{\frac{1}{2}-\mu} f(z) \in L(\gamma - i\infty, \gamma + i\infty),$ $\hfill (4.1.9)$

and as $|z| \to \infty$ in $\overline{S}$,

$$f(z) = o(z^{\mu - \frac{1}{2}}). \qquad (4.1.10)$$

If $|\operatorname{Re} \gamma_0| = a$, then $f(z)$ has to satisfy a Hölder condition in a $\overline{S}$-neighbourhood of γ_0. Then we obtain

$$f(\gamma_0) = -\frac{1}{4\pi i} \int_{-1}^{1} \frac{P_k^{\mu,\gamma_0}(-y)}{(1-y)} \left\{ \int_{\gamma-i\infty}^{\gamma+i\infty} \nu\, \Gamma\!\left(k - \frac{\mu+\nu}{2} + 1\right) \times \right.$$

$$\left. \times \Gamma\!\left(-k - \frac{\mu+\nu}{2}\right) P_k^{\nu,\mu}(y)\, f(\nu)\, d\nu \right\} dy. \qquad (4.1.11)$$

The above expansion formula also holds when $\gamma = \operatorname{Re} \gamma_0 = 0$, and $f(z)$ is only defined on the real line $\operatorname{Re} z = 0$, and satisfies the condition (4.1.8) with

$$\left. \begin{array}{ll} \text{(i)} & z f(z) \in L(0, i), \\[4pt] \text{(ii)} & z^{\frac{1}{2}-\mu} f(z) \in L(i, i\infty), \end{array} \right\} \qquad (4.1.12)$$

and $f(z)$ is of bounded variation in a neighbourhood of $z = \gamma_0$. Then in the left side of the formula (4.1.11), $f(\gamma_0)$ has to be replaced by $\frac{1}{2}[f(\gamma_0 + 0.i) + f(\gamma_0 - 0.i)]$.

Example 4.1.1

Applying the expansion formula (4.1.3) and using the result (1.3.23), for $-1 < x < 1$ we find that

$$(1+x)^{p+1}(1-x)^{k-p-1} = -\frac{2^{k+\frac{1}{2}\mu-1}\,\Gamma(k-p+\frac{1}{2}\mu)\,\Gamma(k-p-\frac{1}{2}\mu)}{2\pi i}\int_{\gamma-i\infty}^{\gamma+i\infty}\nu\,2^{-\frac{1}{2}\nu}\times$$

$$\times\frac{\Gamma(-k-\frac{\mu+\nu}{2})\,\Gamma(p-\frac{1}{2}\nu+1)}{\Gamma(k+\frac{\mu-\nu}{2}+1)\,\Gamma(-p-\frac{1}{2}\nu)}\;\mathrm{P}_k^{\mu,\nu}(x)\,d\nu,\qquad(4.1.13)$$

where p is a complex number, γ is real with $\gamma < \min(2+2\operatorname{Re}p,-\operatorname{Re}(2k+\mu))$, $\operatorname{Re}(k-p) > \frac{1}{4}$ and $|\operatorname{Re}\mu| < 2\operatorname{Re}(k-p)-\frac{1}{2}$.

Example 4.1.2

For $p = -\frac{3}{2}$, the result (4.1.13) becomes

$$(1+x)^{-\frac{1}{2}}(1-x)^{k+\frac{1}{2}} = -\frac{2^{k+\frac{1}{2}\mu}\,\Gamma(k+\frac{1}{2}\mu+\frac{3}{2})\,\Gamma(k-\frac{1}{2}\mu+\frac{3}{2})}{\pi i}\times$$

$$\times\int_{\gamma-i\infty}^{\gamma+i\infty}\frac{\nu\,2^{-\frac{1}{2}\nu}\,\Gamma(-k-\frac{\mu+\nu}{2})}{(\nu^2-1)\,\Gamma(k+\frac{\mu-\nu}{2}+1)}\;\mathrm{P}_k^{\mu,\nu}(x)\,d\nu,\qquad(4.1.14)$$

where $\gamma < \min(-1,-\operatorname{Re}(2k+\mu))$, $\operatorname{Re}k > -\frac{5}{4}$, $-1 < x < 1$.

Example 4.1.3

For $k = -\frac{1}{2}-\frac{1}{2}\mu$, formula (4.1.14) produces

$$(1+x)^{-\frac{1}{2}}(1-x)^{-\frac{1}{2}\mu} = -\frac{\sqrt{2}\,\Gamma(1-\mu)}{2\pi i}\int_{\gamma-i\infty}^{\gamma+i\infty}\frac{\nu\,2^{-\frac{1}{2}\nu}}{(\nu^2-1)}\;\mathrm{P}_{-\frac{\mu+1}{2}}^{\mu,\nu}(x)\,d\nu,\qquad(4.1.15)$$

where $\gamma < -1$, $\operatorname{Re}\mu < \frac{3}{2}$, $-1 < x < 1$.

Example 4.1.4

For $p = -1$, from formula (4.1.13), we find

$$(1-x)^k = \frac{\Gamma(k+\frac{1}{2}\mu+1)\,\Gamma(k-\frac{1}{2}\mu+1)}{\pi i\,2^{1-k-\frac{1}{2}\mu}}\int_{\gamma-i\infty}^{\gamma+i\infty}\frac{2^{-\frac{1}{2}\nu}\,\Gamma(-k-\frac{\mu+\nu}{2})}{\Gamma(k+\frac{\mu-\nu}{2}+1)}\;\mathrm{P}_k^{\mu,\nu}(x)\,d\nu,$$

$$(4.1.16)$$

where $\gamma < -\operatorname{Re}(2k+\mu)$, $\operatorname{Re}k > -\frac{3}{4}$ and $-1 < x < 1$.

Example 4.1.5

For $k = -\frac{1}{2}-\frac{1}{2}\mu$, relation (4.1.16) becomes

$$(1-x)^{-\frac{1+\mu}{2}} = \sqrt{\frac{\pi}{2}}\,\frac{\Gamma(\frac{1}{2}-\mu)}{2\pi i}\int_{\gamma-i\infty}^{\gamma+i\infty}2^{-\frac{1}{2}\nu}\,\mathrm{P}_{-\frac{\mu+1}{2}}^{\mu,\nu}(x)\,d\nu,\quad\operatorname{Re}\mu < \frac{1}{2},\ -1 < x < 1.$$

$$(4.1.17)$$

Example 4.1.6

For $p = -\frac{1}{2}$, the expansion formula (4.1.13) gives

$$\frac{(1+x)^{\frac{1}{2}}}{(1-x)^{\frac{1}{2}-k}} = \frac{\Gamma(k+\frac{1}{2}\mu+\frac{1}{2})\,\Gamma(k-\frac{1}{2}\mu+\frac{1}{2})}{\pi i\,2^{3-k-\frac{1}{2}\mu}} \int_{\gamma-i\infty}^{\gamma+i\infty} \frac{\nu\,2^{-\frac{1}{2}\nu}\,\Gamma(-k-\frac{\mu+\nu}{2})\,P_k^{\mu,\nu}(x)}{\Gamma(k+\frac{\mu-\nu}{2}+1)}\,d\nu,$$

$$(4.1.18)$$

where $\gamma < -\mathrm{Re}(2k+\mu)$, $\mathrm{Re}\,k > -\frac{1}{4}$ and $-1 < x < 1$.

Example 4.1.7

After transforming the term in the left of the integral to the left side of the relation (4.1.18) and then putting $k = -\frac{1+\mu}{2}$, we find that

$$\frac{1}{2\pi i} \int_{\gamma-i\infty}^{\gamma+i\infty} \nu\,2^{-\frac{1}{2}\nu}\,P_{-\frac{\mu+1}{2}}^{\mu,\nu}(x)\,d\nu = 0,$$

$$(4.1.19)$$

for $\mathrm{Re}\,\mu < -\frac{1}{2}$ and $-1 < x < 1$. This result is trivial since the integrand is an odd function of ν.

Example 4.1.8

Applying the integral expansion formula (4.1.3) together with the use of the result (1.3.24), it is found that

$$\sqrt{\frac{1+x}{1-x}} = \frac{-2^{-3-\frac{1}{2}\mu}}{i\,\cos\frac{\pi\mu}{2}} \int_{\gamma-i\infty}^{\gamma+i\infty} \frac{\nu\,2^{-\frac{1}{2}\nu}\,\Gamma(-\frac{k}{2}-\frac{\mu+\nu}{4})\,\Gamma(\frac{k}{2}-\frac{\mu+\nu}{4}+\frac{1}{2})}{\Gamma(-\frac{k}{2}+\frac{\mu-\nu}{4}+\frac{1}{2})\,\Gamma(\frac{k}{2}+\frac{\mu-\nu}{4}+1)}\,P_k^{\mu,\nu}(x)\,d\nu,$$

$$(4.1.20)$$

for $\gamma < \min\{\mathrm{Re}(2k+2-\mu),\ \mathrm{Re}(-2k-\mu)\}$ and $-1 < x < 1$.

Example 4.1.9

Using the expansion formula (4.1.4) and the result (1.3.25), we obtain

$$(1-x)^{k-p}\,(1+x)^p = -\frac{2^{k-\frac{1}{2}\mu-2}\,\Gamma(p-\frac{1}{2}\mu+1)}{\pi i\,\Gamma(-p-\frac{1}{2}\mu)} \int_{\gamma-i\infty}^{\gamma+i\infty} \nu\,2^{\frac{1}{2}\nu}\,\Gamma(k-p+\frac{1}{2}\nu)\times$$

$$\times\frac{\Gamma(k-p-\frac{1}{2}\nu)\,\Gamma(-k-\frac{\mu+\nu}{2})}{\Gamma(k-\frac{\mu-\nu}{2}+1)}\,P_k^{\nu,\mu}(x)\,d\nu,$$

$$(4.1.21)$$

for $\mathrm{Re}\,p > -\frac{1}{4}$, $|\gamma| < 2\,\mathrm{Re}(k-p)$, $\gamma < -\mathrm{Re}(2k+\mu)$ and $-1 < x < 1$.

Example 4.1.10

Putting $p = k - 1$ in the relation (4.1.21), we obtain

$$\frac{(1-x)}{(1+x)^{1-k}} = \frac{i \, 2^{k-\frac{1}{2}\mu-3} \, \Gamma(k-\frac{1}{2}\mu)}{\Gamma(-k-\frac{1}{2}\mu+1)} \int_{\gamma-i\infty}^{\gamma+i\infty} \frac{2^{\frac{1}{2}\nu} \, \nu^2 \, \Gamma(-k-\frac{\mu+\nu}{2}) \, P_k^{\nu,\mu}(x)}{\sin\frac{\pi\nu}{2} \, \Gamma(k-\frac{\mu-\nu}{2}+1)} \, d\nu, \quad (4.1.22)$$

for $\operatorname{Re} k > \frac{3}{4}$, $|\gamma| < 2$, $\gamma < -\operatorname{Re}(2k+\mu)$ and $-1 < x < 1$.

4.2 Integral Expansions in $(1, \infty)$

Theorem 4.2.1

Let c be a real number with

$$c > |\operatorname{Re} \nu| - 2\operatorname{Re} k - 2, \quad (4.2.1)$$

and $f(x)$ a function such that for all $a > 1$,

$$\left.\begin{array}{rll} \text{(i)} & f(x)(x-1)^{\frac{c}{2}-1} \in L(1,a), & \\ \text{(ii)} & f(x)x^{-\frac{5}{4}+|\operatorname{Re}(k+\frac{1}{2})|} \in L(a,\infty), & \operatorname{Re} k \neq -\frac{1}{2}, \\ \text{(iii)} & f(x)x^{-\frac{5}{4}}\ln x \in L(a,\infty), & \operatorname{Re} k = -\frac{1}{2}. \end{array}\right\} \quad (4.2.2)$$

Further, let $f(x)$ be of bounded variation in the open interval $(1, \infty)$. Then, we have

$$\frac{1}{2}[f(x+0) + f(x-0)] = \frac{1}{2\pi i} \int_{c-i\infty}^{c+i\infty} \frac{\mu \, Q_k^{\mu,\nu}(x)}{e^{i\pi\mu}} \left\{ \int_1^\infty \frac{f(y)}{(y-1)} \, P_k^{-\mu,-\nu}(y) \, dy \right\} d\mu, \quad (4.2.3)$$

and

$$\frac{1}{2}[f(x+0) + f(x-0)] = \frac{1}{2\pi i} \int_{c-i\infty}^{c+i\infty} \mu \, P_k^{-\mu,-\nu}(x) \left\{ \int_1^\infty \frac{e^{-i\pi\mu} \, f(y)}{y-1} \, Q_k^{\mu,\nu}(y) \, dy \right\} d\mu. \quad (4.2.4)$$

Theorem 4.2.2

Let c be a real number satisfying the inequality (4.2.1), and c_0 be a complex number with $\operatorname{Re} c_0 \geq 0$ and

$$\operatorname{Re} c_0 > |\operatorname{Re} \nu| - 2\operatorname{Re} k - 2. \quad (4.2.5)$$

Let $f(z)$ be a function continuous on $\operatorname{Re} z \geq \min(c, \operatorname{Re} c_0)$, and analytic on $\operatorname{Re} z > \min(c, \operatorname{Re} c_0)$. Further, let

$$f(z) \, \Gamma(1+z) \in L(c - i\infty, c + i\infty), \quad (4.2.6)$$

115

and as $|z| \to \infty$ on $\mathrm{Re}\, z \geq \min(c, \mathrm{Re}\, c_0)$

$$
\Gamma(1+z)f(z) = \begin{cases}
2^{\frac{3}{2}z}\, z^{-\frac{3}{2}-2k-|\mathrm{Re}\,(2k+1)|}\, o(1), & \mathrm{Re}\, k > -1 \ (k \neq -\tfrac{1}{2}), \\[2mm]
\dfrac{2^{\frac{3}{2}z}}{\sqrt{z}\,\ln z}\, o(1), & k = -\tfrac{1}{2} \text{ and } \mathrm{Re}\, k = -1, \\[2mm]
2^{\frac{3}{2}z}\, z^{\frac{7}{3}+4k}\, o(1), & \mathrm{Re}\, k < -1.
\end{cases}
$$

If $\mathrm{Re}\, c_0 = c \geq 0$, then $f(z)$ has to satisfy a Hölder condition in a right neighbourhood of $z = c_0$. Then $f(z)$ satisfies the following integral expansion:

$$
f(c_0) = \frac{1}{2\pi i} \int_1^\infty \frac{P_k^{-c_0,-\nu}(y)}{y-1} \left\{ \int_{c-i\infty}^{c+i\infty} \mu\, e^{-i\pi\mu}\, Q_k^{\mu,\nu}(y)\, f(\mu)\, d\mu \right\} dy. \tag{4.2.7}
$$

Theorem 4.2.3

Let k be a complex number with $\mathrm{Re}\, k > -1$. Let S be the strip $|\mathrm{Re}\, z| < a$ in the complex z-plane, and $\overline{S}$ the strip $|\mathrm{Re}\, z| \leq a$, where a is a positive number. Let c be a real number and c_0 be a complex number in $\overline{S}$. Suppose $f(z)$ is a function analytic on S and continuous on $\overline{S}$ such that

$$
\frac{2^{\frac{3}{2}z}}{\Gamma(k + \frac{z+\nu}{3} + 1)\, \Gamma(k + \frac{z-\nu}{3} + 1)} f(z) \tag{4.2.8}
$$

is an even function of z in $\overline{S}$. Further, let

$$
\{\Gamma(z)\}^{-1} f(z) \in L(c - i\infty, c + i\infty), \tag{4.2.9}
$$

and as $|z| \to \infty$ in $\overline{S}$

$$
\{\Gamma(z)\}^{-1} f(z) = o(1). \tag{4.2.10}
$$

Let c_0 be a complex number satisfying the inequality (4.2.5) such that $|\mathrm{Re}\, c_o| = a$. Then $f(z)$ has to satisfy a Hölder condition in a $\overline{S}$-neighbourhood of c_0. Then, we have

$$
f(c_0) = \frac{1}{2\pi i} \int_1^\infty \frac{e^{-i\pi c_0} Q_k^{c_0,\nu}(y)}{(y-1)} \left\{ \int_{c-i\infty}^{c+i\infty} \mu\, P_k^{-\mu,-\nu}(y)\, f(\mu)\, d\mu \right\} dy, \tag{4.2.11}
$$

where c is a real number satisfying the inequality (4.2.1).

116

This expansion formula (4.2.11) also holds for $c = \operatorname{Re} c_0 = 0$, and $f(z)$ is defined on the line $\operatorname{Re} z = 0$ satisfying the condition (4.2.8), with

(i) $\qquad zf(z) \in L(0, i)$,

(ii) $\{\Gamma(z)\}^{-1} f(z) \in L(i, i\infty)$.

Also $f(z)$ is of bounded variation in a neighbourhood of $z = c_0$. Thus in the left side of the relation (4.2.11), $f(c_0)$ has to be replaced by $\frac{1}{2}[f(c_0 + 0.i) + f(c_0 - 0.i)]$.

No example illustrating these integral expansions could however be cited here, since we were unable to find an integral similar to (1.3.22) for which the integration range is $(1, \infty)$.

Chapter 5

Some further integral expansions

In this chapter, we present some other kinds of integral expansion formulae involving non-periodic Legendre and associated Lengendre functions $E_{\pm\nu-\frac{1}{2}}(\theta), E_{\pm\nu-\frac{1}{2}}^{-\mu}(\theta)$ where θ is in general complex. For $0 \leq \operatorname{Re} \theta \leq \pi$, these functions can be identified with the conventional Legendre and associated Legendre functions. Before presenting the integral expansion formulae, we mention some properties of these functions relevant to our discussion.

5.1 Some properties of the non-periodic Legendre functions

The function $E_{\nu-\frac{1}{2}}(\theta)$ (cf. Clemmow, 1961) is a non-periodic solution of the Legendre differential equation

$$\frac{d}{d\theta}\left(\sin\theta \, \frac{du}{d\theta}\right) + (\nu^2 - \frac{1}{4})\sin\theta \, u = 0, \tag{5.1.1}$$

for general θ (complex) and is given by

$$E_{\nu-\frac{1}{2}}(\theta) = \sqrt{\frac{\pi}{2}} \, e^{\frac{i\pi}{4}} \frac{\Gamma(\nu+\frac{1}{2})}{\Gamma(\nu+1)} \frac{e^{i\nu\theta}}{\sqrt{\sin\theta}} \, F\left(\frac{1}{2}, \frac{1}{2}; 1+\nu; \frac{-i\,e^{i\theta}}{2\sin\theta}\right), \tag{5.1.2}$$

which for $0 \leq \operatorname{Re} \theta \leq \pi$ is the same as $Q_{\nu-\frac{1}{2}}(\cos\theta - i.0)$ (cf. Erdélyi et al. 1953, p. 146). This has branch points at $\theta = n\pi$ $(n = 0, \pm 1, \pm 2, \ldots)$ in the complex θ-plane. The expression in the right side of relation (5.1.2) can be continued analytically throughout the entire complex θ-plane cut by straight lines running parallel to the negative imaginary axis to infinity from branch points at $\theta = 0, \pm\pi, \pm 2\pi, \ldots$

Another representation of the function $E_{\nu-\frac{1}{2}}(\theta)$ is (cf. Erdélyi et al. 1953, p. 146)

$$E_{\nu-\frac{1}{2}}(\theta) = \frac{\sqrt{\pi}\,\Gamma(\nu+\frac{1}{2})}{\Gamma(1+\nu)} \, e^{i(\nu+\frac{1}{2})\theta} \, F\left(\frac{1}{2}, \frac{1}{2}+\nu; 1+\nu; e^{2i\theta}\right), \tag{5.1.3}$$

where the hypergeometric series for the above hypergeometric function converges for all values of θ on and above the real axis except at the branch points. From the expression (5.1.3), it can be shown that

$$E_{\nu-\frac{1}{2}}(\theta + m\pi) = i^m \, e^{i\pi m\nu} \, E_{\nu-\frac{1}{2}}(\theta), \tag{5.1.4}$$

where $m = 0, \pm 1, \pm 2, \pm 3, \ldots$.

This shows that the function $E_{\nu-\frac{1}{2}}(\theta)$ is non-periodic in θ and the behaviour of $E_{\nu-\frac{1}{2}}(\theta)$ throughout the entire complex θ-plane is specified in terms of its behaviour in the region $0 \leq \mathrm{Re}\,\theta \leq \pi$.

For $0 \leq \mathrm{Re}\,\theta \leq \pi$ (cf. Erdélyi et al. 1953, p. 140),

$$E_{\nu-\frac{1}{2}}(\theta) = \frac{\pi}{2\cos\pi\nu}[i \, e^{i\pi\nu} \, P_{\nu-\frac{1}{2}}(\cos\theta) + P_{\nu-\frac{1}{2}}(-\cos\theta)]. \tag{5.1.5}$$

Since

$$P_{\nu-\frac{1}{2}}(\cos\theta) = -P_{-\nu-\frac{1}{2}}(\cos\theta), \tag{5.1.6}$$

it follows from relation (5.1.5) that

$$E_{-\nu-\frac{1}{2}}(\theta) = \frac{\pi}{2\cos\pi\nu}[i \, e^{-i\pi\nu} \, P_{\nu-\frac{1}{2}}(\cos\theta) + P_{\nu-\frac{1}{2}}(-\cos\theta)]. \tag{5.1.7}$$

From equations (5.1.5) and (5.1.7), we find that

$$E_{\nu-\frac{1}{2}}(\theta) - E_{-\nu-\frac{1}{2}}(\theta) = -\pi \tan\pi\nu \, P_{\nu-\frac{1}{2}}(\cos\theta). \tag{5.1.8}$$

As $\theta \to \pi$ through positive real values, the behaviour of $P_{\nu-\frac{1}{2}}(\cos\theta)$ is wellknown to exhibit a logarithmic singularity (cf. Erdélyi et al. 1953, p. 164). This behaviour shows a logarithmic singularity of $E_{\nu-\frac{1}{2}}(\theta)$ as $\theta \to 0$ or π. Thus, from relation (5.1.4), it is found that

$$\lim_{\theta \to m\pi} \sin\theta \, E'_{\nu-\frac{1}{2}}(\theta) = -e^{i\pi m \, (\nu-\frac{1}{2})}, \tag{5.1.9}$$

where $m = 0, \pm 1, \pm 2, \ldots$ and the dash denotes differentiation with respect to θ.

Using the relations (5.1.5) and (5.1.7), the Wronskian of the functions $E_{\nu-\frac{1}{2}}(\theta)$ and $E_{-\nu-\frac{1}{2}}(\theta)$ is given by

$$W\left\{E_{\nu-\frac{1}{2}}(\theta), E_{-\nu-\frac{1}{2}}(\theta)\right\} = \frac{\pi \tan\pi\nu}{\sin\theta}. \tag{5.1.10}$$

119

This shows that $E_{\nu-\frac{1}{2}}(\theta)$ and $E_{-\nu-\frac{1}{2}}(\theta)$ are two independent solutions of Legendre's differential equation (5.1.1) for $\nu \neq n$ $(n = 0, \pm 1, \pm 2, \ldots)$. From relation (5.1.5), for $0 \leq \operatorname{Re} \theta \leq \pi$

$$E_{n-\frac{1}{2}}(\theta) = E_{-n-\frac{1}{2}}(\theta). \tag{5.1.11}$$

Also from relation (5.1.4), for $m = 0, \pm 1, \pm 2, \ldots$

$$E_{\pm n-\frac{1}{2}}(\theta + m\pi) = i^m (-1)^{mn} E_{\pm n-\frac{1}{2}}(\theta). \tag{5.1.12}$$

Relations (5.1.11) and (5.1.12) imply that

$$E_{n-\frac{1}{2}}(\theta) = E_{-n-\frac{1}{2}}(\theta). \tag{5.1.13}$$

for all θ, if n is an integer.

Another useful result that can be derived from expression (5.1.3) is

$$\lim_{\nu \to \frac{1}{2}} \sin\theta\, E'_{-\nu-\frac{1}{2}}(\theta) = -\cos\theta. \tag{5.1.14}$$

Also

$$\lim_{\nu \to \frac{1}{2}} \left(\nu - \frac{1}{2}\right) E_{-\nu-\frac{1}{2}}(\theta) = -1, \tag{5.1.15}$$

and

$$\lim_{\nu \to \frac{1}{2}} \frac{d}{d\theta}\left\{\sin\theta\, E'_{-\nu-\frac{1}{2}}(\theta)\right\} = \sin\theta, \tag{5.1.16}$$

and

$$\sin\theta\, E'_0(\theta) = -1. \tag{5.1.17}$$

For any fixed value of θ $(\sin\theta \neq 0)$, and any fixed value of $\arg \nu$ in $-\pi < \arg \nu < \pi$, as $|\nu| \to \infty$

$$E_{\nu-\frac{1}{2}}(\theta) \sim \sqrt{\frac{\pi}{2}}\, e^{\frac{1}{4}i\pi}\, \frac{e^{i\nu\theta}}{\sqrt{\nu \sin\theta}}, \tag{5.1.18}$$

where $\sqrt{\nu}$ has a positive real part, and $\sqrt{\sin\theta}$ is positive for real values of θ between 0 and π.

5.2 Some properties of the non-periodic associated Legendre functions

The function $E_{\nu-\frac{1}{2}}^{-\mu}(\theta)$ is a non-periodic solution of the associated Legendre differential equation

$$\frac{d}{d\theta}\left(\sin\theta\,\frac{du}{d\theta}\right) + \sin\theta\left\{\left(\nu^2 - \frac{1}{4}\right) - \frac{\mu^2}{\sin^2\theta}\right\}u = 0, \qquad (5.2.1)$$

for complex values of θ and is defined by

$$E_{\nu-\frac{1}{2}}^{-\mu}(\theta) = \frac{e^{-\frac{1}{2}(\pi\mu-(1+2\nu-2\mu)\theta)}\sqrt{\pi}\,\Gamma(\frac{1}{2}+\nu-\mu)}{2^\mu\,\Gamma(1+\nu)\,\sin^\mu\theta}\,F\left(\frac{1}{2}-\mu, \frac{1}{2}+\nu-\mu; 1+\nu; e^{2i\theta}\right),$$

$$(5.2.2)$$

which for $0 \le \operatorname{Re}\theta \le \pi$ is the same as $Q_{\nu-\frac{1}{2}}^{-\mu}(\cos-i.0)$ (as discussed in Section 5.1). We consider only the case when $\mu > \frac{1}{2}$. Since $a + b - c = -2\mu < 0$, the standard series of $F(a, b; c; e^{2i\theta})$ is absolutely and uniformly convergent for all $\theta \in (-\infty, \infty)$ (cf. Erdélyi et al. 1953, p. 57). Thus $\sin^\mu\theta\,E_{\nu-\frac{1}{2}}^{-\mu}(\theta)$ is a continuous function of $\theta \in (-\infty, \infty)$.

From the relation

$$E_{\nu-\frac{1}{2}}^{-\mu}(\theta + m\pi) = (-1)^{-\mu m}\,e^{i(\frac{1}{2}+\nu+\mu)m\pi}\,E_{\nu-\frac{1}{2}}^{-\mu}(\theta),$$

$m = 0, \pm1, \pm2, \ldots$, it is obvious that the function $E_{\nu-\frac{1}{2}}^{-\mu}(\theta)$ is non-periodic in θ, and the behaviour of $E_{\nu-\frac{1}{2}}^{-\mu}(\theta)$ throughout the entire complex θ-plane is specified in terms of its behaviour in $0 \le \operatorname{Re}\theta \le \pi$ as in the case of $E_{\nu-\frac{1}{2}}(\theta)$.

As $|\nu| \to \infty$, the asymptotic expansion of $E_{\nu-\frac{1}{2}}^{-\mu}(\theta)$ is given by (cf. Erdélyi et al. 1953, p. 136)

$$E_{\nu-\frac{1}{2}}^{-\mu}(\theta) = e^{-i\pi\mu}\sqrt{\frac{\pi}{2}}\,\frac{\Gamma(\frac{1}{2}+\nu-\mu)}{\Gamma(1+\nu)}\,e^{\frac{i\pi}{4}}\,\sin^{-\frac{1}{2}}\theta\,e^{i\nu\theta}\,[1 + O(|\nu|^{-1})].$$

The wronskian of $E_{\nu-\frac{1}{2}}^{-\mu}(\theta)$ and $E_{-\nu-\frac{1}{2}}^{-\mu}(\theta)$ is

$$W\left\{E_{\nu-\frac{1}{2}}^{-\mu}(\theta), E_{-\nu-\frac{1}{2}}^{-\mu}(\theta)\right\} = \frac{\pi\,e^{-2i\pi\mu}}{\sin\theta}\,\frac{\Gamma(\frac{1}{2}+\nu-\mu)}{\Gamma(\frac{1}{2}+\nu+\mu)}\,\frac{\sin\pi\nu}{\cos\pi(\nu+\mu)}. \qquad (5.2.3)$$

From the identities given in Erdélyi et al. 1953, p. 140 and p. 160, it is easy to show that

$$E_{n-\frac{1}{2}}^{-\mu}(\theta) = E_{-n-\frac{1}{2}}^{-\mu}(\theta) \ (n = 0, \pm 1, \pm 2, \ldots), \tag{5.2.4}$$

and

$$\frac{d}{d\theta} E_{\nu-\frac{1}{2}}(\theta) = -i \left(\nu^2 - \frac{1}{4} \right) E_{\nu-\frac{1}{2}}^{-1}(\theta). \tag{5.2.5}$$

Similarly, we can write the following properties of the non-periodic function $E_{\nu-\frac{1}{2}}^{1}(\theta)$:
for $0 \le \mathrm{Re}\ \theta \le \pi$,

$$E_{\nu-\frac{1}{2}}^{1}(\theta) = Q_{\nu-\frac{1}{2}}^{1}(\cos\theta - i.0) = \frac{\sqrt{\pi}\ e^{i(\frac{5\pi}{4}+\nu\theta)}\ \Gamma(\nu+\frac{3}{2})}{\sqrt{2\sin\theta}\ \Gamma(\nu+1)}\ F\left(\frac{3}{2}, -\frac{1}{2}; 1+\nu; \frac{-ie^{i\theta}}{2\sin\theta}\right), \tag{5.2.6}$$

and

$$E_{\nu-\frac{1}{2}}^{1}(\theta) = 2\sqrt{\pi} e^{i(\frac{\pi}{2}+\theta(\nu+\frac{3}{2}))}\ \sin\theta\ \frac{\Gamma(\nu+\frac{3}{2})}{\Gamma(\nu+1)}\ F\left(\frac{3}{2}, \nu+\frac{3}{2}; 1+\nu; e^{2i\theta}\right). \tag{5.2.7}$$

The Wronskian of the functions $E_{\nu-\frac{1}{2}}^{1}(\theta)$ and $E_{-\nu-\frac{1}{2}}^{1}(\theta)$ is given by

$$W\left\{E_{\nu-\frac{1}{2}}^{1}(\theta), E_{-\nu-\frac{1}{2}}^{1}(\theta)\right\} = -\pi \tan\pi\nu\ \frac{(\nu^2 - \frac{1}{4})}{\sin\theta}. \tag{5.2.8}$$

5.3 Integral expansions involving non-periodic Legendre functions

Clemmow (1961) developed formally an integral transform together with its inverse formula involving $E_{\nu-\frac{1}{2}}(\theta)$ and $E_{-\nu-\frac{1}{2}}(\theta)$ where $\theta \in (-\infty, \infty)$, from a unique δ-function representation. The formal construction of the integral transform and its inverse formula is the following.

Equation (5.1.1) can be written as

$$\frac{d}{d\theta}\left(\sin\theta\frac{du}{d\theta}\right) + \lambda\sin\theta\ u = 0, \tag{5.3.1}$$

where

$$\lambda = \nu^2 - \frac{1}{4}. \tag{5.3.2}$$

Then the Green's function $G(\theta, \phi, \lambda)$ satisfies the inhomogeneous equation

$$\frac{d}{d\theta}\left(\sin\theta \, \frac{dG}{d\theta}\right) + \lambda \sin\theta \, G = -\delta(\theta - \phi). \tag{5.3.3}$$

Let $\text{Im}(\nu) > 0$. Then

$$\left.\begin{array}{r}
E_{\nu-\frac{1}{2}}(\theta) \to 0 \text{ as } \theta \to \infty, \\[2mm]
E_{-\nu-\frac{1}{2}}(\theta) \to 0 \text{ as } \theta \to -\infty,
\end{array}\right\} \tag{5.3.4}$$

so that the two functions $E_{\nu-\frac{1}{2}}(\theta)$ and $E_{-\nu-\frac{1}{2}}(\theta)$ are appropriate independent solutions for the construction of the Green's function over the range $-\infty < \theta < \infty$. From relation (1.3), it is found that

$$G(\theta, \phi, \lambda) = \begin{cases}
\dfrac{\cot \pi\nu}{\pi} \, E_{-\nu-\frac{1}{2}}(\theta) \, E_{\nu-\frac{1}{2}}(\phi), & \theta < \phi, \\[4mm]
\dfrac{\cot \pi\nu}{\pi} \, E_{-\nu-\frac{1}{2}}(\phi) E_{\nu-\frac{1}{2}}(\theta), & \theta > \phi.
\end{cases} \tag{5.3.5}$$

The expression (5.3.5) is now used in the representation (1.4), where the contour of integration is closed anticlockwise round all the singularities of $G(\theta, \phi, \lambda)$ in the complex λ-plane. Using the properties discussed in Section 5.1, it is found that the singularities of $G(\theta, \phi, \lambda)$ in the complex λ-plane are a branch point at $\lambda = -\frac{1}{4}$, and simple poles at $\lambda = -\frac{1}{4} + n^2$ $(n = 1, 2, 3, \ldots)$. Since the imaginary part of $\nu = \sqrt{\lambda + \frac{1}{4}}$ is positive, the branch cut is taken along the real λ-axis from $\lambda = -\frac{1}{4}$ to $+\infty$. Then, from the representation (1.4), we obtain for $\theta > \phi$

$$\frac{\delta(\theta - \phi)}{\sin\phi} = -\frac{i}{\pi^2} \int_{-\infty}^{\infty} \nu \cot \pi\nu \, E_{\nu-\frac{1}{2}}(\theta) \, E_{-\nu-\frac{1}{2}}(\phi) \, d\nu, \tag{5.3.6}$$

by changing the variable of integration from λ to ν. The path of integration is to be taken parallel to, but just above, the real ν-axis. The only singularities of the integrand of the representation (5.3.6) in the complex ν-plane are simple poles at $\nu = \pm 1, \pm 2, \pm 3, \ldots$. Then the change in the representation (5.3.6) when the path of integration is translated across the real axis is $\frac{2}{\pi}$ times the sum of the residues at these poles. The residue of the integrand at the pole $\nu = n$ is

$$\frac{n}{\pi} \, E_{-n-\frac{1}{2}}(\phi) \, E_{n-\frac{1}{2}}(\theta), \tag{5.3.7}$$

123

and at the pole $\nu = -n$, the residue is

$$-\frac{n}{\pi} \, \mathrm{E}_{n-\frac{1}{2}}(\phi) \, \mathrm{E}_{-n-\frac{1}{2}}(\theta). \tag{5.3.8}$$

From relation (5.1.13), the expression (5.3.8) is just the negative of the expression (5.3.7), so that the sum of the residues at the poles is zero. Therefore, for $\theta < \phi$, the δ-function representation is obtained from relation (5.3.6) by the interchange of θ and ϕ in the integrand. It is noted that since $\mathrm{E}_{\nu-\frac{1}{2}}(\theta)$ has a logarithmic singularity at its branch point in the complex θ-plane, the right-hand side of relation (5.3.6) is undefined if either θ or ϕ has one of the values $0, \pm\pi, \pm2\pi, \dots$. Hence, a representation of $\delta(\theta - \phi)$ for all real values of θ and ϕ other than $0, \pm\pi, \pm2\pi, \dots$ is obtained by the relation (5.3.6), where the path of integration is indented either above all the poles at $\nu = \pm1, \pm2, \pm3, \dots$ or below them all.

If an arbitrary function $f(\theta)$ is formally written as

$$f(\theta) = \int_{-\infty}^{\infty} \delta(\theta - \phi) \, f(\phi) \, d\phi, \tag{5.3.9}$$

then from the representation (5.3.6), we find that

$$f(\theta) = -\frac{i}{\pi^2} \int_{-\infty}^{\infty} f(\phi) \sin\phi \left\{ \int_{-\infty}^{\infty} \nu \cot\pi\nu \, \mathrm{E}_{\nu-\frac{1}{2}}(\theta) \mathrm{E}_{-\nu-\frac{1}{2}}(\phi) \, d\nu \right\} d\phi. \tag{5.3.10}$$

Changing the order of integration in equation (5.3.10), we obtain

$$f(\theta) = -\frac{i}{\pi^2} \int_{-\infty}^{\infty} \nu \cot\pi\nu \, \mathrm{E}_{\nu-\frac{1}{2}}(\theta) \, F(\nu) \, d\nu, \tag{5.3.11}$$

where

$$F(\nu) = \int_{-\infty}^{\infty} \mathrm{E}_{-\nu-\frac{1}{2}}(\theta) \, \sin\theta \, f(\theta) \, d\theta. \tag{5.3.12}$$

The formula (5.3.12) gives the transform and (5.3.11) is its inverse. Relations (5.3.11) and (5.3.12) together represent the integral expansion of the function $f(\theta)$. Mandal (1973) used these transform formulae (5.3.11) and (5.3.12) to solve a problem of diffraction of waves by a two-part sphere.

5.4 Integral expansions involving non-periodic associated Legendre functions

Using the properties of the non-periodic associated Legendre functions in Section 5.2 and an idea developed by Clemmow (1961) presented in the previous Section 5.3, Maiti (1969) established formally the following integral expansion formula:

$$f(\theta) = -\frac{i}{\pi^2} \int_{-\infty+i\epsilon}^{\infty+i\epsilon} \frac{\nu \cot \pi \nu}{\nu^2 - \frac{1}{4}} \, \mathrm{E}^1_{\nu-\frac{1}{2}}(\theta) \, F(\nu) \, d\nu, \qquad (5.4.1)$$

where

$$F(\nu) = \int_{-\infty}^{\infty} \sin \theta \, \mathrm{E}^1_{-\nu-\frac{1}{2}}(\theta) \, f(\theta) \, d\theta. \qquad (5.4.2)$$

These transform formulae are a generalization of formulae (5.3.11) and (5.3.12) from non-periodic Legendre functions to non-periodic associated Legendre functions. He also applied these transform formulae (5.4.1) and (5.4.2) to solve a boundary value problem of wave propagation in the spherical earth. Mandal (1973) also used these formulae to solve a problem of diffraction of waves by a two-part sphere.

Later, Idemen (1982) established an integral expansion theorem which is a generalization of the transform formulae (5.4.1) and (5.4.2) to the non-periodic associated Legendre functions with superscript $\mu > \frac{1}{2}$. This is given in the form of the following theorem.

Theorem 5.4.1

Let the function $f(\theta)$ be defined for almost every $\theta \in (-\infty, \infty)$, and satisfy the following conditions:

(i) $f(\theta)$ is piecewise continuous and has bounded variation in every finite interval $(-N, N)$,

(ii) $\sin^{1-\mu} \theta \, f(\theta) \in L(-N, N)$ for every $N > 0$,

(iii) $\theta \, f(\theta) \to 0$ as $\theta \to \pm\infty$.

Then for $\mu > \frac{1}{2}$ and $\theta \in (-\infty, \infty)$, we obtain

$$\frac{1}{2}[f(\theta + 0) + f(\theta - 0)] = \int_{L_\infty} \frac{\nu\, F(\nu)}{\sin \pi\nu\, \Gamma(\frac{1}{2} + \nu - \mu)}\, E^{-\mu}_{\nu - \frac{1}{2}}(\theta)\, d\nu, \qquad (5.4.3)$$

where

$$F(\nu) = -\frac{ie^{2i\pi\mu}}{\pi\, \Gamma(\frac{1}{2} - \nu - \mu)} \int_{-\infty}^{\infty} \sin\theta\, E^{-\mu}_{-\nu - \frac{1}{2}}(\theta)\, f(\theta)\, d\theta, \qquad (5.4.4)$$

and L_∞ is the real axis indented above the points $\nu = \pm 1, \pm 2, \ldots$ in the complex ν-plane.

The proof of this theorem is based on the properties of the non-periodic associated Legendre functions $E^{-\mu}_{\nu - \frac{1}{2}}(\theta)$ and $E^{-\mu}_{-\nu - \frac{1}{2}}(\theta)$ discussed in Section 5.2 and the ideas used in Chapters 2 and 3 of this book, so we do not reproduce it.

The integral transform defined by (5.4.4) has some analytical properties similar to Fourier integrals. These can be stated in the form of the following two theorems whose proofs are given in the paper of Idemen (1982).

Theorem 5.4.2

Let us assume that the function $f(\theta)$ satisfies all the conditions of Theorem 5.4.1 together with the following conditions:

(i) $f(\theta) \equiv 0$ for $\theta \in (-\infty, \theta_0)$ with $\theta_0 \in (0, \pi)$,

(ii) $f(\theta) \sim A\, (\theta - \theta_0)^\alpha$ as $\theta \to \theta_0$,

(iii) $\left| f(\theta) \sin^{1-\mu}\theta \right| \leq K \left| \sin^\lambda \theta \right| e^{-c\theta}$ for $\theta_0 \leq \theta_1 \leq \theta$.

Then the transform of $f(\theta)$, $F_-(\nu)$ say, is a regular function of ν in the lower half-plane $\operatorname{Im} \nu < c$, and in this half-plane,

$$F_-(\nu) = O\left(\frac{e^{-i\nu\theta_0}}{\nu^{1+\alpha}\, \Gamma(1 - \nu)} \right)$$

as $\nu \to \infty$. Here $A, \alpha > -1, K > 0, \lambda > -1, c \geq 0$ and $\theta_1 \geq \theta_0$ are suitable constants.

Theorem 5.4.3

Let us assume that the function $f(\theta)$ satisfies all the conditions of Theorem 5.4.1 together with the following conditions:

(i) $f(\theta) \equiv 0$ for $\theta \in (\theta_0, \infty)$ with $\theta_0 \in (0, \pi)$,

(ii) $f(\theta) \sim A\,(\theta_0 - \theta)^\alpha$ as $\theta \to \theta_0$,

(iii) $\left| f(\theta)\sin^{1-\mu}\theta \right| \leq K \left| \sin^\lambda \theta \right| e^{c\theta}$ for $\theta \leq \theta_1 \leq \theta_0$.

Then the transform of $f(\theta)$, $F_+(\nu)$ say, is a regular function of ν in the upper half-plane $\operatorname{Im} \nu > (-c)$, and in this half-plane

$$F_+(\nu) = O\left(\frac{e^{-i\nu\theta_0}}{\nu^{1+\alpha}\,\Gamma(1-\nu)} \right)$$

as $\nu \to \infty$. Here A, $\alpha > -1$, $K > 0$, $\lambda > -1$, $c \geq 0$ and $\theta_1 \leq \theta_0$ are suitable constants.

The integral expansion formula (5.4.3) has been applied in the study of some second-order canonical problems of the geometrical theory of diffraction, namely the diffraction of a whispering gallery mode by a spherical cap by Idemen (1982).

Bibliography

1. Belichenko, V P (1987a), Representation of an arbitrary function in terms of integrals of spherical harmonics. Differential Equations. **23**, 1309 - 1314 (Engl. Transl.).

2. Belichenko, V P (1987b), Diffraction of electromagnectic waves by wedge with anisotropically conducting faces. Zhur. Vychis. Mat. i Mat. Fiz. (Russian). **27**, 889 - 897.

3. Belova, N A and Ufliand, Ia S (1967), Dirichlet problem for a toroidal segment. PMM. **31**, 59 - 63 (Engl. Transl.).

4. Belova, N A and Ufliand, Ia S (1970), Torsion of a truncated hyperboloid. PMM. **34**, 329 - 334 (Engl. Transl.).

5. Braaksma, B L J and Meulenbeld, B (1967), Integral transforms with generalized Legendre functions as kernels. Composito Mathematica. **18**, 235 - 287.

6. Clemmow, P C (1961), An infinite integral transform and its inverse. Proc. Camb. Phil. Soc. **57**, 547 - 567.

7. Erdélyi, A, Magnus, W, Oberhittinger, F and Tricomi, F G (1953), Higher Transcendental Functions, Vol. **1**, McGraw Hill, New York.

8. Erdélyi, A, Magnus, W, Oberhittinger, F and Tricomi, F G (1954), Tables of Integral Transforms, Vol. **2**, McGraw Hill, New York.

9. Felsen, L B (1958), Some new transform theorems involving Legendre functions. J. Math. and Phys. **37**, 188 - 191.

10. Friedmann, B (1956), Principles and Techniques of Applied Mathematics, John Willey and Sons, New York.

11. Gradshteyn, I S and Ryzhik, I M (1980), Table of Integrals, Series, and Products, Academic Press, New York.

12. Hobson, E W (1931), Theory of Spherical and Ellipsoidal Harmonics, Cambridge University Press, London.

13. Idemen, M (1982), On an integral transform with kernel $Q_{\nu-1/2}^{-\mu}(\cos\theta)$ and its application to second order canonical problems of GTD. SIAM J. Appl. Math. **42**, 636 - 652.

14. Jones, D S (1964), The Theory of Electromagnetism, Pergamon Press, New York.

15. Kuipers, L and Meulenbeld, B (1957), On a generalization of Legendre's associated differential equation I and II. Proc. Kon. Ned. Ak. V. Wet. Amsterdam, **60**, 436 - 450.

16. Lebedev, N N (1965), Special Functions and their Applications, Prentice Hall, New Jersy.

17. Lebedev, N N and Skal'skaya, I P (1966a), Integral expansion of an arbitrary function in terms of spherical functions. PMM. **30**, 252 - 258 (Engl. Transl.).

18. Lebedev, N N and Skal'skaya, I P (1966b), Some boundary value problems of mathematical physics and of the theory of elasticity for a hyperboloid of revolution of one sheet. PMM. **30**, 889 - 896 (Engl. Transl.).

19. Lebedev, N N and Skal'skaya, I P (1968), Expansion of an arbitrary function into an integral in terms of associated spherical functions. PMM. **32**, 421 - 427 (Engl. Transl.).

20. Lebedev, N N and Skal'skaya, I P (1986), Integral representation related to Mehler-Fok transformations, Differential Equations. **22**, 1050 - 1056 (Engl. Transl.).

21. Maiti, M (1969), Application of a Legendre function of the second kind in transverse wave propagation in an elastic sphere. Arch. Mech. Stos. **21**, 129 - 143.

22. Mandal, B N (1971a), Note on an integral transform. Bull. Math. de la Soc. Sci. Math. de Roumanie. **15**, 87 - 93.

23. Mandal, B N (1971b), An integral transform associated with the degree of Legendre functions. Bull. Cal. Math. Soc. **63**, 1 - 6.

24. Mandal, B N (1973), Diffraction of waves by a two-part sphere. Indian J. Pure Appl. Math. **4**, 533 - 544.

25. Mandal, B N and Guha Roy, P (1991), A Mehler-Fock type integral transform. Appl. Math. Lett. **4**(4), 29 - 32.

26. Mandal Nanigopal (1995), On a class of dual integral equations involving generalized associated Legendre functions. Indian J. Pure Appl. Math. **26**, 1191 - 1204.

27. Mandal, Nanigopal and Mandal, B N (1992), Integral representation of a function in terms of associated Legendre functions. Bull. Math. de la Soc. Sci. Math. de Roumanie. **36**, 155 - 161.

28. Mandal, Nanigopal and Mandal, B N (1993), Integral expansions related to Mehler-Fock type transforms. Appl. Math. Lett. **6**(1), 17 - 20.

29. Mandal, Nanigopal and Mandal, B N (1994), Expansion of a class of functions into an integral involving associated Legendre functions. Internat. J. Math. & Math. Sci. **17**, 293 - 300.

30. Marcuvitz, N (1951), Field representations in spherically stratified regions. Comm. Pure Appl. Math. **4**, 263 - 315.

31. Moshinskii, A I (1989), Expansions of arbitrary functions in integrals using spherical functions. Differential Equations. **25**, 497 - 501 (Engl. Transl.).

32. Moshinskii, A I (1990), Some integral representations of arbitrary functions in terms of Legendre functions. Differential Equations. **26**, 1179 - 1186 (Engl. Transl.).

33. Nikolaev, B G (1970a), The expansion of an arbitrary function in terms of an integral of associated Legendre functions of the first kind with complex index. Seminar in Math. (V.A. Steklov Math. Inst.), Leningrad. **9**, 45 - 51 (1968).

34. Nikolaev, B G (1970b), Application of an integral transform with generalized Legendre kernel to the solution of integral equations with symmetric kernels. Seminar in Math. (V.A. Steklov Math. Inst.), Leningrad. **9**, 53 - 56 (1968).

35. Pathak, R S (1978), On a class of dual integral equations. Proc. Kon. Ned. Ak. V. Wet., Amsterdam. **81**, 491 - 501.

36. Saxena, R K (1961), A definite integral involving associated Legendre function of the first kind. Proc. Camb. Phil. Soc. **57**, 281 - 283.

37. Sneddon, I N (1972), The Use of Integral Transforms, McGraw Hill, New York.

38. Virchenko, N A (1984), Certain hybrid dual integral equations. Ukr. Math. J. **36**, 125 - 128.